REVOLUTION WITHIN THE REVOLUTION

Revolution Within the Revolution

Cotton Textile Workers and the Mexican Labor Regime,
1910–1923

JEFFREY BORTZ

STANFORD UNIVERSITY PRESS
STANFORD, CALIFORNIA 2008

Stanford University Press
Stanford, California

Library of Congress Cataloging-in-Publication Data

Bortz, Jeff.
 Revolution within the revolution : cotton textile workers and the Mexican
labor regime, 1910–1923 / Jeffrey Bortz.
    p. cm.

Includes bibliographical references and index.

ISBN 978-0-8047-5806-2 (cloth : alk. paper)

1. Textile workers—Mexico—History. 2. Cotton textile industry—Mexico—
History. 3. Labor policy—Mexico—History. I. Title.

HD8039.T42M415 2008
331.88'17721097209041—dc22

                                                                    2007018589

Typeset by Thompson Type in 10/12.5 Palatino

Contents

Tables

Preface and Acknowledgments

I LIVED IN MEXICO CITY for a number of years where I studied wage and labor issues. Over time it became clear to me that Mexican real wages followed many of the rules of underdevelopment: low and with a cycle often determined by trends in the world market. However, the protections and benefits that many workers in the formal sector enjoyed, and the relatively high wages of some industrial workers, ran contrary to some of those rules. After returning to the Unites States, I began to think about this anomaly, which eventually led me to the current study. While working on the problem, I discovered that the origins of Mexican protections and benefits lay in a labor regime created by revolutionary textile workers during Mexico's broader revolution. Subsequent generations of Mexicans benefited from their struggles, so whatever glory there may be in this volume must go to those who made the revolution.

This book argues that the revolutionary labor regime created outcomes favorable to a later political hegemony. The new rules of the work world greatly benefited generations of Mexican workers. Even to date, most Mexican workers enjoy an *aguinaldo*, a product of the workers' revolution. On the other hand, there was no workers' government in Mexico. Common workers did not run the postrevolutionary political system, although union leaders came to enjoy great power. To the degree that corruption and protectionism eventually devoured modern Mexico, and there is quite a debate about this in the literature, rank-and-file workers never constituted a ruling elite, so they can hardly take the blame for these vices, if indeed they were vices. In any case, an analysis of the outcome of the workers' revolution in the 1920s and 1930s is better left to another volume.

Since I did not study history or Latin America as an undergraduate, all of my work in the field owes something to my professors in graduate school at UCLA, Brad Burns, Robert Burr, James Lockhart and Temma Kaplan. Some old friends—Steve Chernack, Dick Dickinson, and Dan Mihaljevich—made sure I got through school and stayed the course thereafter. Zoltan Gross was and is a teacher and a friend, as was Glenda Hubbard later at ASU. Shelby made sure I had access to the obscure materials at the UCLA Library, a place I still enjoy.

I first went to Mexico without knowing much about the country. Marcos Aguila, Francisco Colmenares, Ricardo Pascoe, José Luis Soto, and Edur Velasco lent their assistance and friendship, and to them I owe much of what I learned. José Luis's intricate knowledge of the Mexican work world became a useful starting point for my research. My first job there was at the Mexican Labor Ministry under Porfirio Muñoz Ledo, and I thank him for his support of my work, then and later. I also owe a debt of gratitude that can never be paid to Crescencio Martínez Fernández and Maclovia Soria Juárez, for whom I will always have the greatest affection. I extend this debt to Carmen Juárez, who lived through the revolution and told me many stories, some of which I believed. She is not the author of Chapter 3 on workers, but her spirit is present.

When I eventually returned to the United States, I had to make my way through an academic world that had become foreign to me. Mary Yeager and Stephen Haber facilitated that entry. Steve's work on institutions has made the field much stronger, and his influence on Chapter 6, law, is obvious. At Appalachian State University, Nick Biddle, Larry Bond, and Marv Williamsen kept me focused on the tasks at hand when the distractions threatened to spiral out of control. And there were distractions.

Ma. José Cortés, Quetzalli Cortés, and Itzel Monge assisted me from time to time in the archives. Further assistance was provided by Sandra Mendiola during my stint at the UDLA, Alejandro Martínez Soria in Mexico City, and two graduate students then at ASU, Gregory Swedberg and Sarah Koning. Greg also taught me how to make a good cup of coffee. Mariano Torres introduced me to the archives in Atlixco, and Bernardo García provided assistance in Orizaba. Lisette and Pablo Maurer gave me a place to hang out in Atlixco.

The National Endowment for the Humanities, the Social Science Research Council, a short Fulbright, and the Appalachian State University Research Council provided funding at various stages for the research for this book, and to each I express my gratitude. I want to thank my editor at Stanford, Norris Pope, and also the anonymous reviewer who made this a much better text than it would have been otherwise. An intelligent and accurate critique of the first draft measurably strengthened the argument.

I thank my wife, Josie, for her support through the years in helping me to understand Mexico and other things.

REVOLUTION WITHIN THE REVOLUTION

CHAPTER 1

Introduction

Purpose of the Book

Through most of the twentieth century, Mexico's history differed sharply from the rest of Latin America. When military dictatorships gripped the Southern Cone and dictatorship and revolution swept through Central America, Mexico was an oasis of stable and relatively tolerant, if not exactly, democratic governments.[1] With peace and stability came economic growth, industrialization, and modernization. Without revolution from below or dictatorship from above, Mexico was an island of relative harmony in a Latin American sea of turbulence.

Political harmony is a product of hegemony, which raises the question of what created a hegemonic political system in Mexico. What happened in Mexico that did not happen in other Latin American countries of relatively similar social, economic, and cultural processes? While the obvious answer is the Mexican Revolution of 1910, it is less clear how "the Mexican agrarian revolution," as Frank Tannenbaum aptly named it, could bring lasting peace to a country whose immediate future lay in industry and cities. Mexico needed urban as well as rural peace if it were to emerge from the chaos of revolution. Without doubt, the liquidation of the old land-owning class and the extension of land ownership to millions of campesinos, a process made possible by the rural violence of Emiliano Zapata, Pancho Villa, and other *agraristas*, contributed to postrevolutionary hegemony. By itself, however, land reform could not have produced close to a century of stability in rapidly urbanizing Mexico.

It is a goal of this book to explain how a workers' revolution within the revolution contributed to later political peace in Mexico. From 1910 to 1923, industrial workers challenged authority, threw out the old order, and forced new governments to come to terms with labor. This revolution within the revolution created the most hegemonic, pro-worker labor regime in Latin American history to that point, perhaps to date. It was this labor regime that became the foundation for political hegemony among the social class that represented Mexico's economic and political future, the urban proletariat. Of the many great histories of Mexico's revolution, the one actor ignored by historians has been the winner, the industrial working class. The standard explanation for a new labor relations system has been that it was a gift from above, from the state and its allies in the labor bureaucracy. These explanations either ignore industrial workers or see their participation in the revolution as marginal. As a consequence, the new labor regime appears as the miracle work of politicians and lawyers, a story in which workers do not appear.

Lawyers and politicians, in Mexico as in other countries, were not a particularly generous lot. The generals turned politicians who ruled postrevolutionary Mexico were interested in amassing great sums of money, not limiting the rights of owners to become rich. Carranza was wealthy and Obregón and Calles used their tenures in office to amass as much money as they could. Cárdenas was more modest but not his successors. Why would the wealthy landowners and businessmen who created Mexico's postrevolutionary state want to create an apparently proworker labor regime? It made no sense for them, pursuing their private goals, to have created what must have seemed a workers' paradise in Mexico. While some scholars claim that they carried out their work in order to win elections, Mexico was not an electoral democracy before, during, or after the revolution, so the vote argument simply dissipates.

In fact, the workers' revolution was not a product of the state but rather of workers themselves. Through their actions in the factories and in their communities, mill hands destroyed the old, prerevolutionary labor regime and imposed a new one. While contemporaries acknowledged this,[2] later scholarship did not. In 1976, Ramon Ruiz argued that the activities of workers resulted in a "failure of most of labor to successfully achieve their goals."[3] Twenty-five year later, scholars still claimed that workers benefited from the revolution only because they "encountered revolutionary leaders willing to regulate by decree many of the insecurities of work."[4] This literature did not

- What scholars leave out

explain how industrial workers went from a position of weakness in 1910 to one of strength in 1923 if, as they suggest, the factory was quiet, workers quiescent, and central government nonexistent or weakened for much of that period. It doesn't make sense.

That the state rather than industrial workers shaped the historiography, if not the history, is not surprising. If you look for the state, you will find it. As that knowledgeable bon vivant Salvador Novo remarked during his gay life after the revolution, "In Mexico everything happens according to the spasmodic ejaculations of its politics."[5] Of course Novo was a novelist more attuned to his friends in government than illiterates in the factories. His state, however, did have the capacity to shape historical writing in the decades after the revolution. In 1926, Vicente Lombardo Toledano, the outstanding labor intellectual of the new government, argued that "Union freedom is, in Mexico . . . a new road created *by the State* for the complete emancipation of the proletariat."[6] Later scholars echoed his sentiments, even though the Mexican state from 1910 to 1917, the critical period in the transformation of Mexican labor, could not have created much; it was ceasing to exist.

Later scholars perhaps failed to observe Mexico's workers' revolution because the larger revolution did not follow the classic models of France, the contemporaneous Russian Revolution, or the later Chinese or Cuban revolutions. Most relevant was the contrast with the 1917 Russian Revolution, in which the Bolshevik Party appeared to dominate workers, the revolutionary process, and later, society itself. In contrast, Mexico's revolution lacked, as Alan Knight has argued, "a vanguard party or a coherent ideology."[7] Nonetheless, without either party or a stable set of leaders, Mexican cotton textile workers carried out an antiauthority, antiowner, prolabor revolt from below, employing every means at their disposal from strikes to workplace violence to murder. In the absence of a central government, the workers defeated the owners, imposed unions, a new labor regime, and ultimately union control over the workplace. It was a social revolution from below. Textile owners saw it, but in the midst of the broader rural uprising, it was missed by later scholars.

Since the country's largest and most important factory industry in the early twentieth century was cotton textiles, it is not surprising cotton textile workers were in the forefront of the Mexican workers' revolution. This volume describes the rebellion of cotton textile workers from its outbreak in 1910 to its successful legal conclusion by 1923. While much literature on the revolution has focused on anarchist leaders and heroes, this book contends that it was not them but rather common workers and labor activists who carried out the revolution that was to transform the country's labor regime. This helps explain the deficit in the historiography on Mexico's workers' revolution, because workers who mostly did not write carried out the revolution, not the anarchist leaders and editors of left-wing newspapers.

A second goal of this book is to explain how the workers' revolution in Mexico was fundamentally different from the other great revolutions of the early twentieth century, particularly in Russia. In one area, of course, these revolutions had much in common. Each was a true social revolution, an upheaval from below that fundamentally altered property relations.[8] In Mexico, however, no revolutionary party prepared the revolution, guided workers through it, or took power afterwards. Mexican industrial workers were on their own. This is not to say that workers, artisans, and activists were completely unfamiliar with socialism and anarchism. In fact, anarchists played an important role in founding the Casa del Obrero Mundial and the later Confederación General de Trabajadores (CGT).[9] Marxists founded the Mexican Communist Party (PCM) in 1919, later establishing the Confederación Sindical Unitaria de México.[10] Nonetheless, what is most striking about the workers' revolt in the cotton textile industry is the degree to which it was carried out by activists, militants, and just plain working people, initially angry at specific problems at work or in the community, but not much imbued with socialist or anarchist theory. Without guidelines or a directing party, worker anger and militancy increased in direct proportion to the decline of the capacity of the state to repress them. Radical laborism rather than socialism or anarchism drove Mexico's workers' revolution.

As a consequence, process rather than predetermined plans propelled this revolution. There was no Marx in Mexico, nor Lenin's *State and Revolution*. Without a guiding ideology, the process followed a path that led from concrete work experiences to specific grievances, to generalized complaints, to forming and defending trade unions, and finally to a generalized challenge to the authority of owners. In the beginning, the outbreak of the broader revolution and its impact on government provided an opening for cotton textile workers to express their anger about work: arbitrary and capricious supervisors; unfair disciplinary measures; low wages for long hours; and a generalized understanding that workers were less valued, less important, and less respected than bosses. When they sought a resolution, the owners, supported by long-standing habit, refused to institute changes. In response to the obstinacy of capital, militants and workers created organization because individually they were too vulnerable. Since their historical experience included trade unions, workers in the textile industry quickly turned to them rather than political parties or secret societies.[11] From late 1911, when the revolution's first unions demonstrated their new strength, the process of defending and using trade unions, *sindicatos*, drove the workers' revolution. Owners, who understandably did not want interference with their heretofore unchallenged authority in the factory, tried to fire unionists and destroy unions. Increasingly wide circles of workers responded by defending their organizations at any cost; without unions, there was nothing else they could defend. As conflict in the workplace grew, workers and their unions attacked the authority of foremen, factory administrators, and owners, so that the workers' revolution became a revolution over authority in the factory.

In no other twentieth-century revolution—with the possible exceptions of Bolivia and Spain—did unions play such a central role. Of course, workers' councils, or soviets, were fundamental to the Russian Revolution. In Russia, however, power in the revolution and in the councils fell to the Bolshevik Party. Once in power, the Bolsheviks claimed to rule for workers and their unions but soon suppressed independent union organization.[12] For Mexican textile workers, the vanguard of the workers' revolution, unions were the subject and object of revolution. Trade unions were the main organizational vehicle for workers to challenge owners. This centrality of unions brought success but also limits. The success was that unions replaced owners in controlling the shop floor. The workers' revolt in the cotton textile industry so completely overturned the authority of owners that it transferred control of the factory to unions. Owners controlled the outside—buying machines, materials, and technology, and selling products—but workers' unions controlled the inside—hiring, firing, and disciplining labor. The labor process thus became a process of constant negotiation between managers and union leaders. Unions imposed on owners significant reductions in the hours of work, increases in pay and other benefits, and a new status for common workers. Powerful unions came to control politics in the mill towns and the social mobility of work. In the Latin American context, the gains were many and revolutionary.

The revolution was limited by its relationship to authority. The workers' revolution did not attempt to abolish the authority of the market. Although mill hands disputed the authority of owners to run the factory, they did not dispute ownership or private property. Indeed, trade unions needed all three, owners, private property, and the market, to justify their existence. Therefore, when unions acquired power in the factory, the factory remained subordinate to the market and to private property. Unions quickly discovered that the ultimate limit to the control of work came from outside the factory. Workers' organizations controlled the shop floor but could not determine the price of input materials nor the price of competitors' products. They controlled hiring and firing in the factory but not the supply and demand for jobs in the broader labor market. During the later revolution, workers discovered that they had no alternative but to turn to the state to solve the market problem, to institutionalize their victories, and to protect their unions. If during the early revolution, a collapsing state allowed trade unions to assault owners, during the later revolution trade unions discovered that they needed a state after all. Thus, bowing to the authority of the market meant bowing to the authority of the state, albeit a new, postrevolutionary state.

As a consequence, authority played an ambiguous role during the workers' revolution. Workers challenged authority at work but did not seek to abolish the hierarchy of authority. By disputing the authority of owners rather than authority per se, workers reproduced relationships of hierarchy and subordination in the unions that paralleled the old relationships of hierarchy and subordination in the factories. Similarly, workers did not abolish command, they just moved it from the mill to the union. As a consequence, by the late revolution, union leaders acquired some of the power lost by owners. Authority continued, transferred from mill owners to union leaders. Similarly, although the workers' revolution brought forth some challenges to the authority of men over women, it could no more resolve this problem than that of authority at work. Finally, the workers' revolution represented the victory of modernization, the struggle for rational society and nominally equal citizens. It could not, however, challenge the contradiction of modernity, nominal political equality amidst the fundamental social and economic inequality determined by private property.

Despite the ambiguities, authority played a central role in the Mexican Revolution, as in all social revolutions. William Rosenberg argued that in Russia, "Authority based on traditional social hierarchies in the workplace weakened dramatically."[13] This was because, as McKean and Smith have demonstrated, "working people challenged . . . the all-pervasive authority of factory managers, foremen, and petty workshop proprietors."[14] So too in Mexico, the workers' revolution challenged and weakened authority based on traditional social hierarchies.

Workers achieved their strongest institutional gains in the aftermath of the 1917 Constitution, with state labor codes that ratified a formal shift of authority to workers and their unions. After 1921, however, labor gangsters began a series of labor wars that wracked the textile industry, the communities, and the unions. The outcome did not negate the workers' revolution but did substantially alter its final nature by shifting authority from workers to gangster-run unions. It was a new form of authority but authority over workers nonetheless.

The revolutionary process affected the culture of the working class. The years of violence, of challenging authority, of creating new social relationships at work, altered the attitudes and behaviors of workers toward authority in general, toward traditional hierarchy, and to some extent, toward traditional family and gender roles. The new culture of urban working-class Mexico, later portrayed in film by the great comedians like Tin Tan, reflected and shaped attitudes about race, class, hierarchy, authority, and nation. Many of the traditional ideas about respect for social superiors collapsed during the revolution, often replaced by a culture of violence, of cleverness, and of reliance on friends and family for support at work and in the union. What did not collapse, what the workers' revolution fortified, was respect and love for the nation, another ambiguous participant in the workers' revolution. The workers' revolution was product and cause of developing modernity, as in Russia.

The third goal of this volume is to illustrate how the Mexican workers' revolution shaped the country's institutions well into the present day. A transformation in economic analysis, led by scholars such as Douglass North, Mancur Olsen, and Oliver Williamson, and on Mexico by Stephen Haber, has forced historians and other social scientists to reevaluate the role of institutions in history.[15] This volume argues that the workers' revolution within the revolution completely and thoroughly transformed the formal and informal institutions of Mexico's labor regime. In fact, the workers' revolt not only crushed the old Porfirian labor regime, it gave rise to a new set of institutions, formal and informal, that constituted the most proworker labor regime in the history of Latin America to that time, and perhaps ever. In 1910 there were neither labor laws nor protections for unions, labor organizers, or striking workers. As a consequence, labor organizations in most factories were weak or nonexistent. Owners ran the factories as they wished. They appointed factory administrators who conducted affairs without interference from workers, hiring and firing at will, establishing work rules limited only by custom, and disciplining without any more constraint than a culture that favored rich over poor, white over dark, men over women, and owners (rich, white, and male) over workers (poor, often dark, male and female).[16] Low wages, few benefits, and few workplace rights faithfully reflected the labor market in a country with few jobs in the modern sector and an abundance of unskilled labor.[17]

By 1923, Mexican industrial workers enjoyed the most complete and progressive labor laws in the Western Hemisphere. Article 123 of the 1917 Constitution, state labor codes drafted between 1917 and 1923, and numerous government offices and labor boards protected unions, labor organizers, and striking workers. Indeed, labor law virtually mandated strong unions in most factories. In cotton textiles, owners no longer hired and fired at will. Instead, unions hired and in practice only unions could discharge a worker. Unions and factory administrators jointly negotiated work rules, but only the union could discipline violators. Although custom and culture change more slowly than legal systems, and the older divisions between men and women and blue collar and white collar worker survived, a new culture of solidarity was in place that allowed workers to successfully confront owners. Racism still prevailed, but now many workers expressed a belief in the *raza de bronce*, no longer accepting the automatic superiority of whites and foreigners. Even women earned the right to have other women rather than men direct their unions, obtaining unprecedented equality with men, in law if not in practice.

The institutionalization of the proworker labor regime was almost complete by 1923. In addition to the legal codes mentioned above, its formal components included federal, state, and municipal labor offices, federal and state labor courts, trade unions protected by federal and state law, trade union confederations equally protected, trade union participation in the drafting and enforcing of factory work rules, union participation in the emerging political system, and collective contracts protected by law and courts. These formal institutions continue to operate in Mexico to the present, with the exception of the state labor codes. In August 1931, the federal government substituted a federal labor law that carried forward most of the institutional progress made by workers during their revolution. Meanwhile, the informal components included new attitudes and behaviors of work, solidarity, and authority, as well as new relationships of power and hierarchy in the mills. From nothing to all of this in just a little over a decade is rather impressive by any standards, let alone those of a poor country with a labor market so unfavorable to workers.

The generalization of the new labor regime to the rest of industry and then to the rest of the country mostly took place in the 1920s and 1930s, in part through the 1927 national textile contract and the 1931 Federal Labor Code. In Mexico's most advanced industries, the union shop became the norm, and powerful unions, not businesses, hired, fired, and controlled the shop floor. Workers not only received protective laws, their unions became the dominant political power in working-class communities. Industrial workers received education, health, housing, and employment benefits lacking in the rest of Latin America for the next fifty to one hundred years. Workers in cotton textiles made gains that their counterparts in the United States never achieved, not under the Progressives, not from Roosevelt's New Deal, not even with the post–World War II economic boom.

The existing views of workers in the revolution emanate from backward projection, particularly from the 1940s, when a strong state worked with a powerful labor central to control urban labor. Historians not incorrectly noted that "Labor, both then and later, was generally the subordinate partner in this alliance."[18] True later, it was not the case earlier. The Mexican state that fell apart between 1910 and 1917 did not subordinate anybody; its most outstanding accomplishment was its complete disintegration. The reconstructing state, 1917–23, was so tenuous, so dependent on regional, class, and personal alliances, that it was not in the position to emancipate any social class that had not previously emancipated itself. Even the emerging state, 1923–29, remained dependent on alliances that seriously constrained its activities.

Workers were the primary actors in the creation of the postrevolutionary labor regime. They accomplished their goals in ways that were radically different from the contemporaneous Russian Revolution, yet had a lasting impact on the country's institutions and on the lives of working people. The revolution within the revolution is the missing link in Mexico's modern history.

The Revolution

A brief summary of Mexico's revolution will help the reader better understand the events of the workers' revolution.[19] During the long rule of Spain, 1519–1821, a white, Spanish-speaking elite dominated the Viceroyalty from Mexico City. Although the many different indigenous groups adapted to Spanish ways, particularly through the Church, the province was huge, sparsely populated, regionalized, and only loosely connected. Although silver exports integrated the region into the world market, the mines employed relatively few people, so that most of the population lived and worked in the countryside. An agricultural backwater tied to a Europe-centered world market via silver created the conditions for underdevelopment.

Following Napoleon's invasion of Spain, Miguel Hidalgo initiated the movement for independence in 1810. His call unwittingly unleashed a violent though regional class and race war that thoroughly frightened local elites. With the fall of Napoleon, the Spanish army returned to crush the rebels, though pockets of radicalized resistance continued, particularly in Guerrero in the south. In 1820, the Spanish army at home carried out a coup d'etat to impose a constitutional government. Mexican elites, having fought their own race and class war, were in no mood to tolerate popular rule. They supported Agustín de Iturbide, the head of the Royalist army in New Spain, when he declared a reactionary independence in 1821 to forestall the consequences of modernity and democracy.

Though Emperor Iturbide was quickly overthrown, for seventy years no government established stability or legitimacy. Elites were profoundly divided on how to modernize Mexico, whose underdevelopment became increasingly clear to them. Between 1821 and 1876 the government changed hands seventy-five times.[20] Meanwhile, wealthy whites continued to rule over mostly poor, dark agriculturalists. Mining continued to dominate the economy but still failed to provide enough good jobs to help most Mexicans emerge from poverty. Central America left the Mexican Republic following the Iturbide coup; the United States invaded and took Texas, California, and the sparsely inhabited north in 1846; the French invaded and imposed an emperor in 1860; and liberal and conservative elites used the French disaster to unleash a civil war. During the first half century of independence, the Republic of Mexico faired poorly.

In 1876, the liberal hero of the war against the French intervention, General Porfirio Díaz, took power in a coup that appeared to be just another of Mexico's many military adventures. Instead, he imposed the longest period of political stability in Mexico's independent history. Ruling until 1911, his government coincided with the late-nineteenth-century boom in the world economy. An able negotiator, Díaz allowed foreign investment to prosper. New wealth produced economic growth without generating enough jobs in the modern sector to solve poverty and underdevelopment. To the contrary, the great estates, the haciendas, grew at the expense of the small Indian and campesino villages, a recipe for rural unrest and declining incomes. A few families owned millions of acres while more and more rural communities saw a decrease in their small holdings and a decline in their standard of living. Local and foreign elites disdained the social problem, refusing to believe that concentration of land ownership, spurred by the new railroads, might cause social unrest. In 1907 Nevin Winter claimed "that some seven thousand families practically own the entire landed estate of the country does not inspire envy in the bosoms of the other millions."[21] Two years later, Reginald Enock commented that "Mexican society falls into lines of marked class distinction. The rich and the educated stand in sharp juxtaposition to the great bulk of poor and uneducated, and the high silk hat and frock coat form a striking contrast to the half-naked and sandaled peon in the plazas and streets of the cities."[22] Porfirian elites ignored the poor in their midst, and more dangerously, the growth of regional economies and regional elites isolated from the central government.

When the aging Díaz decided to run for reelection in 1910, a wealthy landowner from Coahuila, Francisco I. Madero, challenged him, claiming that Díaz was a dictator who carried out dishonest elections. In response, Díaz jailed Madero and handily won the June 1910 elections. Madero escaped from jail and on November 20, 1910, issued the Plan de San Luis Potosí, calling on the country to rise up in revolution against the dictator. Madero's call to revolution is now celebrated as the official beginning of Mexico's revolution.

Although Madero's revolution envisaged political reform for the literate classes, in the state of Morelos, popular leader Emiliano Zapata rebelled against the Díaz government, seized large estates, and began to divide the land among poor communities. In Chihuahua, Pascual Orozco and a former bandit, Pancho Villa, led a rebellion against local Porfirian authorities. Against Madero's orders, Orozco and Villa attacked and defeated the federal army at Ciudad Juárez, May 10, 1911. Both Madero and Díaz were frightened by a popular revolt that appeared to be out of control. They negotiated the surrender of the Díaz government on May 21 with the Treaty of Ciudad Juárez. On May 25 Díaz resigned the presidency, leaving the Secretary of Foreign Relations, Francisco León de la Barra, as interim president of Mexico.

When Díaz sailed for Paris, he and Madero believed that the revolution had ended. Madero planned to disarm his supporters, run for election to the presidency, which he would no doubt win, and install the political reforms for which he launched the revolution. In June he met with Zapata to order him to lay down his arms. Zapata refused, demanding that the haciendas return lands that had been illegally stolen from Indian and campesino communities. The federal army, still under the command of the old Porfirian general Victoriano Huerta, invaded Morelos. When local villagers resisted, a separate revolution was underway. The poor villagers in the south demanded a radical land reform rather than the timid political changes of Madero.

On October 1, 1911, Madero won probably the freest elections in Mexican history, taking office on November 6. While he expected a return to Porfirian peace, deposing Díaz opened the door to all the social unrest that had accumulated since the defeat of the popular insurrection of 1810, combining race, class, nationalism, the land problem, underdevelopment, poverty, and modernization. On November 28, Zapata announced the Plan de Ayala, whose goal was the restoration of hacienda lands to the villages. The Plan recognized the revolutionary leadership of Pascual Orozco instead of Madero. Within a week, on December 2, cotton textile workers launched the first general strike in Mexican history, for higher wages, lower hours of work, and to settle other workplace grievances. Soon other strikes by industrial and rural workers followed. Shortly thereafter, Bernardo Reyes, former Porfirian army leader, launched a counterrevolution in the north, after which the controversial Orozco also entered into rebellion against Madero. By October 1912, the revolution was in disarray. The old elites believed, as did José Fernández Rojas, that Mexico's lower classes were a "mixture of black and Indian . . . a terrible human subspecies" unleashed by Madero and bent on revenge.[23] They saw the northern rebels like Orozco and Villa as a lower class with "a revolutionary movement stamped with the socialist doctrines of the XV century."[24]

Madero hoped for a peaceful transition that would open the door to politics for people like himself, wealthy regional outsiders. Instead, the old owning classes hated him for unleashing the subalterns. The campesinado in Morelos hated him for not resolving the land question. Erstwhile allies in the north hated him because he supported the old ruling groups. Madero had few allies and his days were numbered. The timid president refused to change the Porfirian leadership of the army, believing that only the military protected him. Instead, in February 1913, Victoriano Huerta captured Madero and his vice president, José Pino Suárez. After a meeting with Henry Lane Wilson, the U.S. ambassador to Mexico, Huerta had both Madero and Pino Suárez assassinated on February 22, 1913, in front of Lecumberri Prison. He then made himself president in order to restore Porfirian peace.

The next day, the wealthy landowning governor of Coahuila, Venustiano Carranza, an ardent Madero supporter, announced that he would oppose the unconstitutional government of Huerta. He organized the Constitutionalist military resistance, with himself as first chief. Pancho Villa organized the Constitutionalist army in Chihuahua, while Alvaro Obregón led the Constitutionalist forces in Sonora. These men launched Mexico's second revolution, characterized by "the increasing radicalism of all factions."[25] For the next year and a half, extreme violence gripped the country.

Carranza's Constitutionalists were really three separate armies, Villa's Division del Norte in Chihuahua, Obregón's army in the northwest, and another army in the northeast. Obregón and Villa were particularly effective but the cooperation between them and with Carranza was pragmatic at best, nonexistent at worst. Meanwhile, Zapata continued in rebellion in the south, without allying to the Constitutionalists whom he did not trust. Adding to the mix, in April 1914, Woodrow Wilson ordered the U.S. military to capture Veracruz, depriving Huerta of his principal source of revenue and the ability to pay his troops. In June, Villa captured Zacatecas, forcing Huerta to flee the country in July. Many thought that the Maderista revolution had succeeded at last and that Mexico's days of violence were over. Instead, the defeat of Huerta initiated "the most chaotic [years] in Mexican revolutionary history."[26] As John Womack argued, "The struggle within the Mexican regime to restore its constitutionality had resulted in its destruction—the collapse of all the labyrinthine national, regional and local political and business deals developed over the previous 30 years."[27] It also destroyed the old federal army, mainstay of political power and the state since its founding by Iturbide.

Huerta abandoned Mexico in a German ship, leaving his foreign minister, Francisco Carbajal, as interim president in Mexico City. The victorious revolutionaries, really separate armies, leaders, and programs, met in October at the Convention in Aguascalientes to determine a government. This was a conflictive meeting; Zapata's representatives argued that Madero's revolutionary program, effective voting and no reelection, should give way to the Zapatista revolutionary program of land and liberty. Zapata and Villa eventually broke with Carranza and Obregón. The Convention supported the former and appointed Eulalio Gutiérrez president of Mexico, with Villa the head of its army. It declared Carranza in rebellion. Carranza, still first chief of the Constitutionalists, moved his government to the textile center of Orizaba, entrusting the Constitutionalist forces to Alvaro Obregón. By December 1914, Mexico had two governments, Gutiérrez and the Convention in Mexico City and Carranza and the Constitutionalists in Orizaba. The country now entered into its third revolution, further radicalizing all parties. Because the old ruling classes and their army had been decisively defeated, this ensured that one faction or another of revolutionaries would win power.

In early January 1915, Gutiérrez fled Mexico City, and on January 28, Obregón occupied it and imposed Constitutionalist rule. Villa reorganized a government in the north while Zapata continued entrenched in Morelos. The Constitutionalists took the offensive, defeating Villa at Celaya in April. In desperation, Villa ordered an attack on Columbus, New Mexico, March 1916, provoking another U.S. invasion when John J. Pershing led a group of young troops that included Douglas McArthur, Dwight D. Eisenhower, George S. Patton, and others in pursuit of Villa. They withdrew in January 1917 without capturing Villa.

By late 1916, Carranza's Constitutionalists dominated much of the country. It was a critical moment because the old regime was dead, many Constitutionalists had radicalized since Madero, and the Villistas and Zapatistas were weakened but dangerous enemies. Carranza needed a new Constitution to legitimate his revolution and his victory. A Constitutional Convention met in Querétaro from November 1916 to February 1917, issuing its new Constitution that month. Radicals far to the left of the conservative Carranza dominated the convention. Although the new Constitution closely followed its 1857 predecessor in structure, three Articles transformed it into a revolutionary code. Article 3 mandated free public education and state control of educational affairs, significant in a country in which the Church had once dominated that realm. Article 27 provided the legal basis for land reform and made private property a privilege subordinate to the needs of the people and the state. Finally, Article 123 became the most extensive and radical labor reform in Latin American constitutional history to that time.

With the Constitution in place, the victorious Constitutionalists organized elections in March 1917 that made Carranza president. He took office in May, and the next year ordered General Pablo González to conduct a military campaign in Morelos to defeat Zapata. In April 1919, Colonel Jesús Guajardo assassinated Zapata, effectively ending the long Zapatista rebellion, alienating many agraristas and some

Constitutionalists. Nonetheless, assassination was an effective tool that the regime continued to employ. In July 1920, a defeated Villa accepted a government offer to retire to his ranch in Durango. Obregón had him assassinated in 1923; a similar fate met other former Obregón allies like Lucio Blanco, abducted in the United States and assassinated in Mexico, and Benjamin Hill, the minister of war who was poisoned.

When Carranza took power, his key allies were the military leaders from Sonora, Alvaro Obregón, Plutarco Elías Calles, and Adolfo de la Huerta. These men came to power as leaders of the Constitutionalist military venture in the northwest. They rose to the top because of their success in organizing the military, thus it was natural for them to solve problems through the method they knew best, killing. When Carranza attempted to name his successor in the 1920 presidential elections, the three of them announced a rebellion. Carranza fled Mexico City, only to be assassinated by a follower of Obregón, who assumed the presidency (1920–24). When Obregón attempted to name his successor (Calles, 1924–28), de la Huerta launched a rebellion that was defeated shortly thereafter. Subsequently, Obregón ran for reelection to the presidency to succeed Calles, leading to the revolt of former allies, Francisco Serrano and Arnulfo Gomez, whom he defeated and executed shortly before falling to the assassin's bullet himself in July 1928.

Calles survived as the strongman of Mexico's revolution. He realized that the revolution had assassinated its leaders, Madero, Villa, Zapata, Carranza, Blanco, Hill, Serrano, Gómez, and even Obregón. To avoid this fate, in 1929 he organized the Partido Nacional Revolucionario (PNR), which occupied the presidency and controlled the state from 1929 until 2000.

At the center of the Revolution, from Díaz to the PNR, stood the state. Madero declared a revolution against the head of state, Huerta assassinated the head of state, Carranza wrote a new Constitution to ratify his state, and Calles founded a party of the state. The state is the embodiment of the social relations of class, which is why the assault on the state in Mexico quickly led to a social revolution of the lower classes.

The pillar of nineteenth- and twentieth-century Mexico's market economy was private property, a social construct defined and enforced by the state. Embedded in the concept of private property is the right to appropriate someone's labor. The appropriation of labor, as well as goods and services, is a function of legal rules, organizational forms, and social norms and behaviors. The labor regime is the set of social relationships, organizations, and institutions that define the appropriation of labor and its products, the parameters of workplace rules, concepts, and behaviors. These parameters determine what Gordon, Reich, and Edwards label the "socially expected organization of the labor process."[28] It includes the formal and institutional labor relations system and also the informal but even stronger constraints of attitudes, cultures, and behaviors of work and workplace relationships.

Within the broader Mexican Revolution, the workers' revolution is better understood through the concept of labor regime rather than moral economy. Most historians have used moral economy to understand "how economic activities of all kinds are influenced and structured by moral dispositions and norms, and how in turn those norms may be compromised, overridden or reinforced by economic pressures."[29] Some of the best works, like *Weapons of the Weak* by anthropologist James C. Scott, have argued that hegemony is always contested, the resistance of the weak rooted in values that lie outside the norms of the dominant economy.[30]

Clearly the workers' revolution in Mexico, part of the larger and complex Mexican Revolution, expressed a conflict between a powerful social group, mill owners, and those who felt weak and exploited, mill workers. However, the concept of labor regime addresses this clash most directly by exploring the nature of a disputed workplace and the institutional arrangements, formal and informal, through which conflict and resolution took place. The workers' revolution was about the social relations of work and must be understood in reference to contested authority *at work.*

There is a complex interplay between the norms and rules that allow people to appropriate labor and the norms and rules that govern the social relations of work. It is in the interest of property holders in industry to extend the meaning of property to control over the workplace: hiring, firing, and disciplining labor, and defining and controlling the work process. It is equally in the immediate interest of working people to maximize their own control over the work environment. "Employees' interests conflict with those of their employers even at the moment that they produce goods and services and not just at the end of the day when the fruits of their productive activity are distributed amongst the different factors of production."[31] In every society, the outcome of this conflict/cooperation between those who work and those who control work establishes the parameters of the social relations of work. The formal and informal rules of work affect the meaning of private property and are a fundamental component of the state as a relationship between social classes.

In Mexico, the state and the workplace had rested on the authority of property. The revolution began by challenging the head of state and ended by destroying the state. With long-standing grievances, cotton textile workers took advantage of its collapse to challenge the authority of property. That challenge led to the destruction of the old labor regime in Mexico and prior definitions of private property rights. The new labor regime, built on a compromise between workers, unions, and owners, required a new system of authority and of property. The new authority and the new labor regime became an integral and necessary construct of Mexico's new state. That is why the workers' revolution played such an important role in Mexico's revolutionary and postrevolutionary outcomes.

Plan of the Book

This book follows the logic of the workers' revolution, beginning with a description of the textile industry and its workers, tracing the opening of the revolution and the transformation of the challenge to authority, moving to the institutional victories in law and trade unions, and concluding with a view of how the new system operated in the early institutional period.

With 142 mills and 33,000 workers in 1907, cotton textiles was Mexico's largest, most important, and most advanced factory industry. A workers' revolution, to be successful, would have to take hold of this industry or fail. Transport, rail, mine, and oil workers labored in industries whose working conditions varied too greatly from the factory norm to serve as a model for future labor relations. Cotton textile factories, however, could become models for newer industries, such as steel and automobiles, that would shape Mexico's later industrialization. Chapter 2, "The Mexican Cotton Textile Industry," presents a brief history of the industry from its establishment in the 1830s through its turmoil during the revolution. It describes the companies, factories, mill towns, textile zones, and production processes that became the location of Mexico's workers' revolution. Of particular importance was the geography of the industry, a single corridor from Mexico City through Puebla and Atlixco to Orizaba, which undoubtedly potentiated the workers' rebellion.

Cotton textile workers lived in textile communities, mill towns, factory housing, and working-class neighborhoods that facilitated an easy communication among them. Chapter 3 describes the "Layered Communities of Mexican Cotton Textile Workers," which is organized in three areas: social class, social relationships, and social ideas. Not surprisingly, for textile workers, their ideas of revolution emerged from their experiences on the shop floor and in their communities.

Chapter 4 portrays "The Beginning of the Workers' Revolution, 1910–1912." Madero's challenge to Díaz led to a rapid unionization of the largest and most important mills in 1910–11. Unionization and the weakening of central authority brought the first general strike in Mexican history in December 1911. In a historic moment, textile workers successfully shut down the industry. The government responded in July 1912 by convening a meeting of textile industrialists to solve the industry's labor problems. At the meeting, industrialists and government officials drafted the first industrywide collective contract in Mexican history, although they refrained from labeling the documents a contract. Convinced they had won the general strike, textile workers continued to pursue changes in the mills, carrying out important strikes and labor actions throughout 1912 and 1913.

During this period, labor actions increasingly attacked the authority of owners, leading to Chapter 5, "Challenging Authority, 1912–1916." This chapter describes how workers tried to resolve their grievances in the early revolution by challenging the traditional authority of owners and managers, how the weakness of central government emboldened their challenges, and how the extension of that challenge led to unprecedented demands for worker control in the factory. As the central government collapsed, regional military commanders in the textile zones tried to pacify workers through military decrees that legislated labor affairs. The chapter concludes with a description of the decrees and their impact on the new consciousness of hierarchy and authority in the mills.

The collapse of central authority in Mexico, the unionization of the mills, and the consistent workers' challenge to authority ultimately destroyed the old labor regime. The military decrees demonstrated to new elites that it was necessary to institutionalize labor affairs, the subject of Chapter 6, "The Institutionalization of the Labor Regime: Law and Government." It begins with the first institutionalization, the convention and protocontract that culminated the 1911 general strike. With the military decrees of 1914–16, it became obvious that the Constitutionalists would draft a labor code, which became Article 123 of the 1917 Constitution. This Article built on the military decrees, a direct response to the workers' revolt in cotton textiles. Conflict at the Constitutional Convention left the states to implement Article 123 through state labor codes. The most important were those in the textile states of Veracruz (1918) and Puebla (1921). Chapter 6 concludes with an analysis of the labor codes in the two states.

Trade unions were the primary vehicle for the workers' revolution and the subject of Chapter 7, "The Institutionalization of the Labor Regime: Unions." During the early revolution, mill hands fought heroically to create and defend organization. This chapter studies the nature of the early organizations as well as the changes that took place through the revolution. It also describes the growth of textile unions, the relationships of union leaders to rank and file, and some of the limits to a union-based workers' revolution. Although workers did not ignore local political organizations and elections, there is little doubt that they believed that trade unions most completely represented their goals. The result was a steadily increasing role for unions throughout the revolution, until they became dominant. Law, studied in the previous chapter, reinforced the power of unions, and together they constituted the major institutional victories of the workers' revolution.

Chapter 8, "Labor Conflict in the Early Institutional Period, 1917–1923," describes the changes in the early institutional period after the passage of the 1917 Constitution. It analyzes the increase in labor violence, the changed trade unions of that period, and most important, the new power of workers and unions over the shop floor. The labor regime of 1917–23 had little in common with that of the Porfiriato.

Chapter 9, "The Revolution and the Labor Regime, 1910–1923," concludes by demonstrating that workers won their revolution through an analysis of the changes in wages, hours, and benefits, and also control over hiring, firing, and discipline. It also analyzes the limits to the workers' victory, rooted in its focus on law and trade unions, both of which required a strong state. A revolution born of a collapsing state ended up by creating a new one, with all its attendant ambiguities.

Conclusion

In 1910, Mexico erupted in revolution. Initiated by regional elites who felt excluded from central power, the intraelite squabble opened the door to a massive social rebellion from below. In a country that was largely rural and agrarian, it is not surprising that agrarian revolutionaries like

Emiliano Zapata and Pancho Villa became the foremost spokesmen for the campesinado's principal demands, land and liberty.

Meanwhile, the urban proletariat, led by workers in the country's largest and most important factory industry, cotton textiles, launched their own revolution within the revolution. Distant from Europe, their revolution was fundamentally different from that of contemporaneous Russian workers, who replaced the Tsarist state with a workers' state. Mexican workers did not focus on the state but instead made the workplace a site of permanent conflict between workers and management. They focused on building and sustaining trade unions and using those unions to obtain worker control over the shop floor.

This revolution within the revolution generated a radical change in Mexico's labor regime, the "socially expected organization of the labor process."[32] By assaulting the owners' control over the factories, workers and their unions forced emerging elites to institutionalize a fundamentally new set of social relationships of work. Between the early military decrees of 1914 and 1915, a new Constitution in 1917, and state labor codes between 1918 and 1921, Mexico created the most hegemonic, proworker labor regime in Latin American history. The outcome of this struggle is described in this book's conclusion.

Committed to an industrial and urban future, Mexico's postrevolutionary elites used the new labor regime to construct the most hegemonic political system in Latin America to that point. The system may not have been democratic, it may not have lifted Mexican economic performance above the norms for Latin America, but it did provide a peaceful transition to a new society and economy. The labor regime that emerged after the revolution may have favored unions over workers, but it provided industrial workers with greater benefits than those achieved in other Latin American countries at the time, and greater benefits than would have emerged otherwise from an underdeveloped labor market. The workers' revolution within the revolution did not create a perfect world, but for those who made it, it created a better world and one fundamentally different from its prerevolutionary counterpart.

CHAPTER 2

The Mexican Cotton Textile Industry

THIS CHAPTER DESCRIBES THE MEXICAN cotton textile industry from its establishment in the 1830s through its turmoil during the revolution. The companies, factories, mill towns, textile zones, and production processes were the site and object of Mexico's workers' revolution. This chapter will highlight four elements that most strongly influenced the shape of that revolution. First was the spectacular growth of an industry that constituted Mexico's industrial revolution, for without that there would not have been a revolution of industrial workers. Second, the development of ownership groups with strong foreign participation gave workers the possibility of combining their social goals with a national/patriotic discourse, which strengthened their feelings and cohesion. Workers constantly complained about the foreigners in the industry. Third, the concentration of industry in four industrial corridors immeasurably facilitated communication among rebellious workers, and ultimately, unionization. The geographical compactness of the cotton textile industry was a critical factor in facilitating labor organizing and revolution. Finally, the profitability and importance of the industry in early twentieth-century Mexico allowed it to survive and function during the years of upheaval. Had the factories shut down, there would have been no workers' revolution. But most did not close and workers continued to work, creating space for their activities.

A Brief History of the Industry up to the Revolution

As R. J. Peake noted more than a century ago, "The arts of Spinning and Weaving are among the oldest in the world."[1] Arabs brought its manufacture to Iberia; the English and Spanish "cotton/algodon" derive from the Arabic, *qutun*.[2] After 1760, the British transformed the cotton trade and then the entire world with the Industrial Revolution, which emerged from their manufacture of cotton textiles. It became a model of development for the rest of the world, first in the advanced countries and later in the backward countries.[3]

In Britain, modern trade unions and factory strife grew alongside the new industrial system. In other countries that industrialized through cotton textiles, unionization and factory conflict often followed. Outside of western Europe, the American, Indian, and Russian textile industries were particularly successful in economic terms, but they also caused their share of labor strife. In the United States, textiles were home to some of the country's most virulent labor conflicts, from the women's-led Lowell, Massachusetts, strike in 1834, through the Industrial Workers of the World-led strikes just before World War I, to the violent, nationwide walkout in 1934. Like the U.S. industry, Russian industry was "overwhelmingly financed by native capital" and by 1913 "was Russia's largest industrial employer . . . with the 690,124 workers in 1,449 textile enterprises accounting for 29 percent of the country's factory labor force."[4] In 1917, Russian textile operatives were at the forefront of the world's first self-proclaimed workers' revolution, one that challenged traditional hierarchy and authority in the factory in the midst of a wider revolution. Led by their Soviets, textile workers abolished private industry in textiles. The textile industry also had its radicals in India, where in 1920 Mahatma Gandhi founded the Textile Labour Association.[5]

Cotton was a principal item of clothing in Mesoamerica prior to the Spanish conquest. In Cortes's first letter to Charles V, written from Veracruz in 1519, he commented that "the women of rank wear skirts of very thin cotton."[6] After settling in Mexico, the Spanish were not interested in the native cloth, importing their own from Europe. Through the long colonial rule (1519–1821), the upper classes continued to import cloth and clothes. Nonetheless, indigenous producers carried on with the production of cotton textiles, utilizing what Spaniards called *telares sueltos* or scattered looms. Spaniards also developed local production of cloth in artisan centers, *obrajes*, with both forced and free labor. The number of both types of labor varied greatly through the years. For the early eighteenth century, scholars have estimated there were forty-nine obrajes in Mexico City with 2,200 workers and thirty-two obrajes in Puebla with 1,440 workers.[7] Most obrajes developed near population centers, with a majority in Mexico City, Puebla/Tlax-cala, and the Bajío.[8] Many produced wool, which had been unknown in Mexico before the Spanish conquest. As a consequence, the technology for wool production was "imported whole cloth, so to speak, from late medieval Europe and experienced little innovation before the end of the eighteenth century."[9] Others produced cotton and cotton products. By 1793, there were an estimated 7,800 telares sueltos and thirty-nine obrajes in New Spain.[10]

Pedro Sáinz de Baranda, Lucas Alamán, and Esteban de Antuñano brought industrialization to Mexico in the 1830s. In 1830, Sáinz de Baranda became Jefe Político in Valladolid, where he soon established Mexico's first mechanized textile factory, La Aurora Yucateca. Within ten years, La Aurora employed 117 workers, a modest success until it was destroyed by the Guerra de las Castas.[11] However it was really the Banco de Avío, a development bank founded in 1830 by Lucas Alamán, a preeminent figure of early independent Mexico and a minister in Anastasio Bustamante's government, that sparked Mexico's industrial revolution and the modern cotton textile industry. The purpose of the bank was to finance industrialization with capital provided by tariffs on cotton textile imports. It underwrote the purchase of foreign textile machinery while subsidizing foreign technicians to travel to Mexico to teach local weavers to operate them."[12]

Born in Veracruz in 1792, Esteban de Antuñano was raised in Spain, then traveled to England where he witnessed firsthand the industrialization of cotton textiles. Returning to Mexico, he married Bárbara Avalos, daughter of a Galiciano who owned the hacienda of San Mateo in Atlixco, which decades later would be used for a textile mill.[13] Antuñano used his properties to finance the purchase of a wheat mill, the hacienda and molino of Santo Domingo. With financing from the Banco de Avío, Antuñano imported machines and contracted ten *maestros ingleses*. He opened La Constancia Mexicana, which many consider the first modern textile operation in Mexico, in Puebla on January 7, 1835.[14] La Constancia spun crude cotton and sold the yarn to artisans who transformed it into cloth.[15] He then built La Economía on the Río Atoyac in Puebla.[16]

In 1836 Lucas Alamán entered into a business agreement with the Legrand brothers, together building a textile factory, Cocolapam, in Orizaba (Veracruz).[17] Cocolapam became the country's largest mill in the first half of the nineteenth century. Meanwhile, Gumersindo Saviñón soon transformed El Mayorazgo from flour to cotton production.[18] These

TABLE 2.1

Mexican Cotton Textile Industry, 1843–1927

Year	Number of factories	Number of spindles	Number of looms	Number of workers
1843	59	106,708	2,609	n.d.
1862	57	133,122	n.d.	n.d.
1898	118	468,574	13,944	21,960
1907	142	694,000	23,500	33,000
1917	99	573,072	20,489	22,187
1927	132	777,380	29,290	41,008

SOURCE: Fernando Pruneda and Miguel A. Quintana, "Estudio sobre la modernización de la industria nacional textil de algodón . . . ," STyPS, Mexico City, April 1943, 37, in AGN, Archivo Gonzalo N. Robles, Caja 76, Exp. 9; For 1907, C. Reginald Enock, *Mexico, It's Ancient and Modern Civilisation History and Politcal Conditions Topography and Natural Resources Industries and General Development* (London, 1909), 337.

early entrepreneurs followed the same path, converting flour mills into cotton mills, utilizing nearby rivers for power, and importing English machines and technicians.

The Banco de Avío established a fund of one million pesos to provide capital for the industry. Despite political turmoil and civil wars that prevented the bank from functioning optimally, five collections of cotton machines and one of wool were eventually sent to Mexico City, Puebla, Morelia, Celaya, Tlaxcala, and Querétaro, the colonial centers of cotton production.[19] The first factories transformed raw cotton into a crude fiber that was then sold to traditional weavers, who made cloth by hand.[20] Antuñano expressed the dream of early industrialists and other progressive Mexicans, "The modern machines established in Mexico and worked by Mexicans, will occupy our weavers with more advantage than the present looms . . . they will develop our cotton agriculture to an incredible degree . . . and they will bring together in the country many millions of pesos, whose circulation will make prosper all the branches [of industry] . . . besides, the modern machines established here, will cheapen the price of textiles, and with that, alleviate [the condition of] the poor classes, as is just and necessary."[21]

Many poor Mexicans bought the cheap cotton cloth, or manta, produced by the factories even though the very wealthy preferred imported cloth; the poorest Mexicans, often subsistence agriculturalists in isolated communities, generally lacked the money to buy much of anything. The local cloth was less expensive than imports because of the newly established government tariffs and de facto protection of international transport costs. With a modest local market and initial financing by the Banco de Avio, Mexican capitalists imported machines, built large factories, and brought foreign technicians to oversee the production process.[22] The creation of a cotton textile industry and a true factory system was Mexico's industrial revolution, although one limited to a single industry and based on imported rather than locally developed technology.

The industry grew steadily, from eight mills in 1836 to seventeen in 1840, forty-seven in 1843, fifty-two in 1844, and sixty-two in 1845.[23] These modern factories were much larger than the old workshops. Saviñón's El Mayorazgo had 2,400 spindles, Antuñano's La Constancia Mexicana had 7,680, and his La Economía Mexicana had 3,900.[24] The factories demanded larger and more secure supplies of raw cotton than the old obrajes, which helped stimulate a shift in the countryside from small ranchos to larger agricultural units. The mechanized cotton textile industry thus contributed to the growth of haciendas and the concentration of land ownership.[25] In the factories, not only production, but also total factory productivity, labor productivity, and capital-labor ratios increased significantly through the nineteenth century.[26]

Mexico's industrial revolution differed considerably from England's. In England, inventors like Henry Cort, James Hargreave, Richard Arkwright, and Samuel Crompton invented, developed, and refined the technology that substituted machines for people.[27] High productivity made their products competitive in domestic and foreign markets. In Mexico, capitalists imported their machinery and technology and often the most skilled workers. The technology was modern but not cutting edge, and it was never truly a function of local conditions. It was simply what foreigners had to offer, and this never changed. When Labor Inspector Miguel Casas visited the large Río Grande mill in Jalisco in 1913, he noted "As can be seen in the manufacturers brands each machine is of English origin and from the best manufacturers known in England."[28]

In Mexico, markets and water power determined factory location. With high transport costs, it was advantageous to locate factories near the country's main urban markets. Additionally, the factories used water power to drive their machines; with the later transition to electrical power, the factories' turbines also required rivers. Urban markets and rivers explain why the industry developed near Mexico City and Puebla, the two largest markets, and Atlixco and Orizaba, near high volcanoes and flowing rivers, one town near Puebla and the other on the road between Puebla and the port of Veracruz. The concentration of factories in a few areas created mill towns and textile zones in a virtual textile corridor.

Many of the textile factories were built in old haciendas with water mills to grind wheat. The new establishments acquired some of the characteristics of the old haciendas: large, ornate buildings, workers' houses on the premises, paternalism in running the enterprise, a male-oriented work culture, and a chapel, or at the very least, a space for the Virgin of Guadalupe, an expression of the society's profound Catholicism. In 1877, for example, the administrator of Cocolapam, Alberto G. Morphy (sic) ordered the construction of a chapel dedicated to the Virgin of Guadalupe.[29]

During Mexico's first half century of independence, widespread violence, constant political instability, numerous foreign invasions, and persistent lack of capital plagued the country's economic growth. Aspiring industrialists suffered from high costs, a lack of native technology, financing difficulties, poor transport systems, an inability to export, and an anemic internal market. In light of these obstacles, the success of the early cotton textile factories was quite surprising, a testimony to the ingenuity of owners and workers alike.

As technologies became more capital intensive in Europe and the United States, Mexico adopted these innovations despite the abundance of unskilled workers. Although this increased the industry's productivity, it still did not make Mexican factories competitive with foreign mills. Productivity in the Mexican industry was barely sufficient to allow Mexican producers to compete in local markets; the country's factories never acquired the capacity to compete in foreign markets. Mexico's new industry imported technology and exported little, mostly producing cheap cloth for local markets. In 1862 the country's 133,122 spindles in fifty-seven factories produced 7,853,779 pounds of yarn (the specialty of twelve mills) and 1,258,963 pieces of cloth (the specialty of forty-five mills).[30]

In 1876, General Porfirio Díaz (1876–1910) took power in a successful military uprising and surprised the country by imposing the longest period of political stability in Mexico's independent history.[31] Buoyed by a stable government and a subsequent upswing in international commerce, the Mexican economy experienced its first period of sustained economic growth, which between 1893 and 1907, "grew more rapidly than that of the United States."[32] Investments in railroads and the consequent decline in transport costs benefited the national economy and the textile industry, which needed to ship cotton from long distances and send its products throughout the large country. The railroad dropped the cost of shipping one ton of cotton from Mexico City to Querétaro from 61 pesos in 1877 to 3 pesos in 1910.[33] The industry grew modestly from 1876 to 1890, then expanded strongly during the 1890s as companies, factories, and machines became larger and more modern. From 86 mills, 8,132 looms, 234,386 spindles, and 10,871 workers in 1877, productive capacity jumped to 141 mills, 18,553 looms, 600,707 spindles, and 28,192 workers in 1900.[34]

During the late Porfiriato, the industry underwent horizontal and vertical integration, even greater capital-intensive production, and ownership consolidation.[35] New companies with larger and more modern factories came to dominate the industry. From the individually owned private companies of the mid-nineteenth century, there was a turn to limited liability joint stock companies, with some giant companies publicly traded on the new Mexico City stock exchange. The principal firms, the Compañía Industrial de Orizaba (CIDOSA), the Compañía Industrial Veracruzana (CIVSA), the Compañía Industrial de San Antonio Abad (CISAASA), and the Compañía Industrial de Atlixco (CIASA), "were mammoth enterprises, even by US standards."[36]

Outside of mining, cotton textile manufacturing "was the largest industry in Mexico, accounting for almost one-quarter of the total value of manufactured goods produced in the Republic."[37] It was an industry that had long depended on foreigners, and this did not change during the Porfiriato. Many American and English engineers and technicians participated in the construction of the factories and the installation of modern machinery and up-to-date hydroelectric plants. Meanwhile, an influx of Spanish (Puebla) and French (Orizaba and Mexico City) industrialists changed the ownership groups. Foreigners dominated because Porfirian textile factories were closely linked to the establishment of large commercial establishments in the country's main cities and their ties to banking groups, areas in which the foreigners and their friends played important roles.

Cotton manufacture stimulated the commercial production of the plant, contributing to the transformation of the rural hacienda. The industrial zones that manufactured cloth and products were mostly quite far from the regions that grew cotton. The Comarca Lagunera, centered on the city of Torreon, was the country's principal cotton producer, and its importance grew during the Porfiriato. The region covers 2,760 square kilometers in Coahuila and Durango. Between 1910 and 1920

TABLE 2.2

Mexican Cotton Harvest, 1909–1927 (kilograms)

1909	36,061,200	1916	31,752,000	1922	43,697,882
1910	36,061,200	1917	29,484,000	1923	38,024,938
1911	35,380,800	1918	30,318,000	1924	42,568,500
1912	31,752,000	1919	47,628,000	1925	43,467,029
1913	34,020,000	1920	48,762,000	1926	78,016,452
1914	28,350,000	1921	33,339,600	1927	38,862,252
1915	28,350,000				

SOURCE: Juan Chavez Orozco, *Monografía Económico-Industrial de la Fabricación de Hilados y Tejidos de Algodon* (Mexico, 1933), p. 22.

the Mexicali Valley also began to produce cotton, becoming the second most important growing region in the country. By 1930 there were some 70,000 hectares under cultivation.[38] Both the Lagunera and Mexicali are very dry zones in which agriculture depends on irrigation, the latter from the Colorado River.

By the end of the Porfiriato, cotton textile production had become a mature industry in Mexico. Per capita consumption of textile products did not compare unfavorably with the poorer European countries. Although there were many privately owned small factories, these now coexisted with huge factories, many owned by publicly traded joint stock companies with significant investment by foreigners. With modern machinery and hydroelectric plants, the factories made wide use of the new railroad system to bring in raw cotton and to send finished products to market. In 1909 the British engineer C. Reginold Enock noted that Mexican "cotton mills are amongst the foremost in the world, and their large capacity and splendidly-built factories are a source of surprise to the European or American traveller."[39]

Owners and Ownership

As we have seen, the growth of the industry brought important changes in owners and ownership groups. Antuñano established an initial pattern, later followed by other pioneers of the industry. He was born in Mexico, had European ties and experience, and married into a landed family. Pedro Sáinz de Barranda, who founded the first mill in the Yucatán, was born in Campeche. His parents sent him to Spain to study. After returning to Mexico, he used his political connections to establish his mill, la Aurora.[40] Lucas Alamán was born in Guanajuato, and his parents also sent him to Europe to study. The three men came from comfortable families. As the industry expanded, most industrialists were sons of wealthy families with close ties to Europe and often to landholding families. They were not, however, a homogeneous group, and often they were divided by nationality, family, and economic ties.

Strong European connections continued to characterize ownership in the early twentieth century, though the reasons changed. As the industry grew, "European enterprise and capital played a large part; French and Spanish immigrants were responsible for the principal new plants of that period. Some of these immigrants started in the retail business, dealing in imported fabrics. Then they branched out into manufacturing by setting up mills to supply their own stores with the light-weight cotton goods for which there was such a large popular demand in Mexico . . . this linkage between distribution and manufacturing has continued . . . All the larger department and dry goods stores in Mexico City also own and operate cotton textile mills. The industry is thoroughly integrated along vertical lines."[41]

The consequence was that ownership was divided among Mexicans and immigrants from Spain and France. The Spanish families, mostly in Puebla, owned medium and small mills, as did most of the Mexican owners. The French holdings included the Compañía Industrial de Orizaba (CIDOSA), Cía Industrial Veracruzana (CIVSA), Cía Industrial de San Ildefonso, Cía Industrial de Guadalajara, Cía Industrial de Atlixco (CIASA), Cía Industrial de la Teja, La Perfeccionada, and La Abeja.[42] The Spanish families dominated Puebla; the French dominated Orizaba; and Mexico City ownership included Mexicans, Spaniards, and French.

In 1900 there were 147 textile owners in Puebla: 98 Spaniards, 38 Mexicans, 9 French, and 2 others.[43] The Spaniards mostly came from the industrialized north of their country, the French from the Barcelonnette valley. In Atlixco, Lions Hermanos y Cía ran El Léon from 1898 to 1911, then Signoret y Renaud took control.[44] According to Gamboa, the Spanish textile families included Artasánchez, Conde y Conde, Díaz Rubín, Gavito Méndez, González Cosio, Rivero Quijano, and Sánchez Gavito. Conde y Conde and Gavito also included a Mexican branch. Lion and Rebattu were the principal French families.[45]

The career of José Rivero Quijano in Puebla was somewhat typical of the Spanish owners. An important industrialist during the revolution, he later wrote an impressive amateur history of the industry. He owned El Mayorzago, which began life as a flour mill in the sixteenth century, property of the regidor of Puebla. In 1604 Diego de Carmona aquired the mill and adjacent land through various *mercedes de tierra*. In 1839 a descendent, Joaquin de Haro y Tamariz sold the property to Gumersindo Savigñon, who sold it to Rivero Quijano's great grand-father, José Quijano de la Portilla, in 1864. Quijano de la Portilla had emigrated from Spain to Mexico in the early nineteenth century to work in the family business headed by his aunt, Soledad Manero Vda de Portero. His daughter, Carmen, married her cousin, Alejandro Quijano, who subsequently purchased El Mayorazgo, a "finca de campo," a factory, and a mill. Alejandro converted the flour mill into a textile operation, San José Mayorazgo, taking advantage of water power from the adjacent Río Atoyac. In 1894 José Rivero Quijano closed the flour mill in order to concentrate on cotton. The mill's cotton ouput increased slowly until the 1890s, then more rapidly thereafter. The family also came to control two other small mills, El Molino de Enmedio and San Juan de Amandi.[46]

Despite the growth of his family business, Rivero Quijano was a small businessman competing against increasingly large firms. In 1919, the four largest companies controlled 37.5 percent of the market.[47] In 1921, the largest mills employed thousands of workers: Río Blanco and Santa Rosa in Orizaba employed 2,592 and 1,554 respectively; Metepec in Atlixco 1,162; Covadonga in Puebla 997; La Carolina in Mexico City 1,048; and El Hércules in Querétaro 1,115. Between 1878 and 1910, the median number of spindles per mill doubled to 4,488.[48] In the most modern plants, hydroelectric power and automatic spindles replaced water power and human-powered machinery. By 1900, 44 percent of Mexico's hydroelectric power was generated by plants installed in textile factories.[49] In spite of the changes, however, Mexican industrialists still did not carry out significant research and development, so productivity lagged behind Europe and the industry continued to import its machines.[50]

The French-owned CIDOSA was the giant of the industry with four large, modern mills in Orizaba. It bought Cerritos and San Lorenzo in 1889, opened Río Blanco in 1892, and brought Cocolapam into the fold in 1899. Before 1890, the average mill had only 300 spindles, but Río Blanco opened with 35,000 spindles and 900 looms. Another giant, the Compañía Industrial Veracruzana, started the Santa Rosa mill near Orizaba in 1898 with 33,000 spindles and 1,400 looms. Near Atlixco, French capital opened Metepec in 1902 with 36,852 spindles and 1,570

TABLE 2.3
TABLE 2.3

Estimates of Mexican Cotton Textile Production, 1909–1927

Year	Value, real output	Meters of output	Year	Value, real output	Meters of output
1909	43,370	314,228	1919	23,333	305,509
1910	50,651	315,322	1920	27,840	298,829
1911	51,348	341,441	1921	66,826	338,346
1912	72,834	319,668	1922	53,040	330,601
1913	33,978	298,897	1923	44,214	303,090
1914			1924	44,155	285,594
1915			1925	56,839	380,041
1916			1926	60,562	327,487
1917	12,266		1927	51,156	308,940
1918	15,111	180,453			

SOURCE: Stephen Haber and Armando Razo, "Political Instability and Economic Performance: Evidence from Revolutionary Mexico, *World Politics*, Vol. 51, No. 1, October 1998, 123.

looms. By the end of the Porfiriato, these large, modern factories with automatic looms and electric spindles dominated the industry.[51]

A subtle geographic shift accompanied the increasing size of textile companies and factories. Mexico City and Puebla city continued to be important textile centers, but Orizaba and Atlixco emerged as genuine mill towns, the driving force of the industry, with societies and economies dominated by the massive factories. Older textile centers in Querétaro and Jalisco continued to produce, and their factories became larger, but their share of national ouput declined.

While large firms and mills dominated the industry, the smaller mills did not go out of business. Many smaller enterprises, employing from one hundred to three hundred workers, and some less than fifty, continued to operate.[52] The large mills usually carried out a variety of activities, from spinning and weaving cloth to making cotton products. The smaller plants often specialized in a particular activity. A 1927 government survey divided the industry into ninety-nine spinning and weaving mills, which were the core of the industry: nineteen spinning; thirteen weaving; eight spinning and knitting; four spinning, weaving, and knitting; twelve spinning, weaving, and stamping; three stamping; and one dyeing.

During the early years of the twentieth century, textile owners organized two competing groups. Adrian Reynaud, B. D. Salceda, H. Gerard, Juan N. Nieto, Emilio Meyran, Antonio Donnadiu, and Francisco

Martinez Arauna founded the Centro Industrial Mexicano (CIM) in early 1907 on the heels of and because of labor conflict in 1906.[53] With a powerful base among Puebla and Tlaxcala industrialists, the CIM was nonetheless not very effective in confronting the 1911/1912 general strike. Many industrialists wanted a newer, more aggressive organization, and turned to leaders of the old political class to organize one. Chief among them was Tomás Reyes Retana. Reyes Retana had served as a federal congressman in the 1890s and later as a Porfirian senator representing Mexico City. A member of various legislatures with Porfirio Díaz, Francisco León de la Barra, and Francisco Madero, he was keenly aware of the changes in the labor climate. On May 8, 1912, Reyes Retana founded the Confederación Fabril Nacional Mexicana (CFNM). Among the original members of this group were Antonio Reynaud of CIDOSA, Félix Vinatier of the Compañía Industrial Veracruzana, Leon Barbaroux of La Hormiga, and other powerful industrialists or lawyers mostly outside of the Puebla/Tlaxcala group that now dominated the CIM. The early CFNM represented twenty textile factories, mostly large, modern mills, seven each in Mexico City and Veracruz, four in Jalisco, and one each in Querétaro and Guanajuato.[54] The CIM continued to defend the interests of the small and medium-sized factories of Puebla and Tlaxcala.

The Industry During the Revolution

Even during the last years of Porfirio Díaz, foreign and national elites believed that he "enjoyed the confidence of his fellow citizens and a majority of their votes."[55] His governments favored and supported industrialists, and the textile industry experienced significant growth and change during his time in office. The outbreak of revolution in 1910 surprised even the revolutionaries.

During that period, the growth of the largest factories was closely tied to the growth of Mexico's largest banks. They were connected industries that were relatively profitable, despite the bank crisis that struck Mexico in 1907. Although many scholars have attributed the growth of industry during the late nineteenth century to Porfirian political peace and stability, recent scholarship has demonstrated that the instability of revolutionary Mexico from 1910 to 1929 had less of an impact than one might imagine.[56] From 1910 to 1913, the changes in government from Díaz to León de la Barra to Madero to Huerta had little effect on the economy. From 1913 to 1917, during the worst of the violence, the destruction of the transport system and the spreading warfare caused an important downturn in economic activity, but the recovery was rapid and continued into the 1920s. What is clear is that nineteenth-century industrialists built an important and profitable industry that produced goods for which there was an ample market. Even during the worst of the revolution, the industry continued to function.

There is some evidence that cotton textile factories may have suffered a bit less than other industries. In December 1914 the Labor Office reported that industry had suffered from the revolution but that cotton textiles "Is that which has suffered the least . . . because the majority of its products are sold in nacional markets."[57] There are other reasons that the industry continued to function: the revolution was only sporadically violent, some regions suffered more greatly than others, and most revolutionaries had little interest in shutting down the factories. Although there were local supply and transportation difficulties, the industry continued to produce basic goods that found a market among the war-ravaged population.

Nonetheless, the cotton textile industry, like the rest of the industry, experienced revolution. In September 1913, El Hércules closed "for lack of raw materials, which have not been able to enter the Republic for the interruption of railroad traffic."[58] In November 1913, fourteen Puebla mills temporarily shut down "for lack of cotton."[59] The Madero family itself owned a cotton factory, La Estrella, and the assassination of Francisco Madero shut down the mill, its "workers have risen in arms since the fall of the Maderista government."[60]

Between 1910 and 1917, the number of registered working cotton textile factories declined only a fifth, while the number of employed workers dropped by a similar amount. The evidence for the most violent years, 1913 to 1917, is contradictory. Although some authors have claimed that many factories shut down, the archives suggest that most continued working. The impact of the violence was unequal and undoubtedly some mills, like Metepec, operated sporadically between 1914 and 1919.[61]

The Mexican revolution was highly regionalized so that each region had a unique experience. Mexico's main textile zones lay in the center of the country's revolution: Mexico City, the Puebla-Tlaxcala valley, and the mountains of Veracruz. The Maderista revolution broke out in Puebla. When Carranza fled Mexico City, he established his provisional government in Veracruz. From neighboring Morelos, Zapatistas attacked and at times controlled Mexico City and Atlixco. In the latter city, a hostile newspaper claimed that "the Zapatista hordes, during their stay in Atlixco, dedicated themselves to theft and looting," reporting the murder of 62 Constitutionalist soldiers.[62] Another paper noted that when Constitutionalists recaptured the town, they captured three prisoners in Tochimilco, who were *"pasados por las armas."*[63] When federal labor inspectors traveled to San Martín Texmelucan, Puebla, to resolve a textile strike in 1914, they not only found workers engaged in a wildcat action to remove a supervisor, they discovered that Zapatistas had captured the road back to Mexico City, blocking their return.[64] When Carranza fled Mexico City in 1920, he took the train to Puebla, ending with his assassination in Tlaxcala. Sudden violence, murder, social upheaval, and rapid political shifts plagued the country's main textile centers throughout this period. The lack of a stable state and a decade of persistent violence affected social relations more than the economic indicators might suggest.

Although Mexico remained a violent and uncertain country throughout the 1920s, the economy recovered sharply through most of the decade. Gross domestic product (GDP) increased 12.1 percent from 1921 to 1927, and the volume of manufacturing production jumped 71 percent.[65] By the late 1920s, cotton textiles was, as it had been two decades earlier, Mexico's most important factory industry. In 1927, Ernest Gruening stated that, "Textiles are Mexico's leading manufactures. One hundred and thirty-five cotton mills employ 44,250 workers, of whom 33,412 are men, 7,424 women, and 3,414, minors. Some 60 per cent. of the industry is French-owned, 35 per cent. Spanish; and British, American, and Mexican capital has one or two mills each."[66] Only the electrical power industry represented a greater investment, while the second largest factory industry, tobacco, represented only a fourth of the money invested in textiles.[67]

Textile Zones and Factories

Mexico of the revolution was a vast, heterogenous country. It is, even today, the world's eleventh largest nation. Two mountain ranges, the Sierra Madre Occidental and the Sierra Madre Oriental, run north-south, roughly from the U.S. border to Mexico City, where they converge and continue south to Central America. The ranges make much of the country a series of mountain valleys or plateaus, arid or semi arid in the north, with difficult communication between them and to the rest of the country. In 1910, the Maya of the Selva Lacandona had

TABLE 2.4

Mexico

Cotton Textile Factories, 1910–1927

Year	Number of establishments	Looms	Spindles	Workers	Cotton consumed	Crude cloth produced (tons)	Yarn (tons)
1910	127	26,184	723,963		35,169	9,587	
1917	99	20,489	573,092	22,187	15,315	6,092	
1918	104	25,017	689,173	27,680	17,482	6,954	
1919	114	27,020	749,237	33,185	27,664	11,004	
1920	120	27,301	753,837	37,936	31,694	10,839	
1921	121	28,409	770,945	38,227	35,924	12,286	
1922	120	27,819	758,624	39,677	34,654	11,851	
1923	113	27,770	752,255	38,684	32,344	10,778	2,602
1924	109	26,536	721,580	37,080	30,517	8,880	2,819
1925	124	28,934	780,691	42,359	40,997	13,466	3,825
1926	131	29,446	786,144	44,114	41,523	12,760	3,456
1927	132	29,290	777,380	41,008	41,170	13,204	3,358

SOURCE: INEGI, *Estadísticas Históricas de México* (Mexico, 1994), 616–617. For 1917–1922, Secretaría de la Economía Nacional, Departamento de Estudios Económicos, *La Industria Textil en Mexico, El Problema Obrero y Los Problemas Económicos* (México, 1934), 14.

little in common with the tall, Spanish-speaking military colonists of Chihuahua, and even less with the entrepreneur-farmers of Sonora who still fought Indian wars. Few of these groups had much sympathy for Mexico City, nor did *capitalinos* look upon *provincianos* as equals. Before the railroad, Mexico was hardly a country at all. Even with the railroad, the many languages and cultures made national integration a formidable task. Mexico was a country of regions, a difficult challenge for any federal government.

Equally regional was the textile business. The first industrialists located their mills near markets, water power, and the old textile centers. This led to an inevitable concentration of cotton textile production, with 80 percent of the industry in a single corridor that stretched from Mexico City east through the neighboring state of Mexico to the Puebla/Tlaxcala valley, then just over the border into the state of Veracruz and the mountain municipios neighboring Orizaba. Although there were scattered mills in twenty-three states from Nuevo Léon (4) and Coahuila (8) in the north to the Yucatán (1) in the south, of the 188 registered factories in 1921, 52 were in Puebla, 31 in Mexico City, 11 each in Veracruz and the State of Mexico, and 10 each in Hidalgo and Tlaxcala.[68] The eleven active mills in Veracruz contained almost 24 percent of the country's total capital invested in the industry. Of the industry's 77,121,849

TABLE 2.5

Regional Production of Cotton Textiles, 1880–1910

State	% of National Production	
	before 1880	in 1910
Puebla	23%	32%
DF	14%	12%
Veracruz	9%	21%
Tlaxcala	4%	6%
Others	50%	29%

SOURCE: Dawn Keremitsis, *La Industría Textil Mexicana en el Siglo XIX* (Mexico, 1973), 123.

pesos in fixed investments, the 56 Puebla mills occupied first place with 25,890,824, the 11 Veracruz mills second with 18,850,847, and the 27 Mexico City factories third, with 9,997,737, with other states holding lesser amounts.[69]

The concentration of the industry into textile zones contributed to close communication among textile workers and to the creation of proletarian culture and later, a culture of revolution. Mexico City, Puebla, Atlixco, and Orizaba had the most mills, the most workers, and produced the most product. Mexico City was too large and too diverse to become a fully proletarian city, although its working class weighed heavily on the city's life. The state capital, Puebla, was also quite diverse, but a concentration of industry lent weight to the struggles of industrial workers. Atlixco and Orizaba, on the other hand, became genuine mill towns in which millworkers dominated numerically, culturally, and later, politically.

Mexico City

Mexico City lies in the valley of Mexico, "a broad elevated plain, or basin, surrounded by hills, which culminate far away to the south-east in the snow-clad summits of Popocatepetl and Ixtaccihuatl, the extinct volcanoes of the Sierra Madre."[70] Built on the foundations of the old Tenochtitlan, the city of the Mexica immediately became the locus of Spanish life during the colonial period. In 1824, the country's new Constitution created the Federal District, which initially included Mexico City. Between 1895 and 1900 the District incorporated formerly independent municipios, Coyocacán, Tacubaya, Tlalpan, San Ángel, Milpa Alta, and Xochimilco.[71] Economic ties extended the city into surrounding municipios in the Valley of Mexico that belonged to the State of Mexico, particularly Naucalpan, Huixquilucan, and Tlalnepantla. The city was, as always, "the political, the educational, the social and the commerical centre of the whole country."[72]

The beautiful colonial center of the city was "surrounded on all sides, except that leading to Chapultepec, by miles of squalid streets, where dwell the poor and outcast of the community—and their name is legion."[73] Although the newly incorporated outlying areas remained well connected to the city, traveling there often meant a long journey through fields and agricultural villages. Many seemed like the distant villages of *provincia*, quite unlike the city proper. The District's population grew steadily during the revolution, from 720,753 in 1910 to 906,063 in 1921 to 1,229,576 in 1930.[74]

From colonial times, Mexico City was a center of artisan and pre-industrial manufacturing. In 1794, the artisan population of the city constituted 29 percent of the laboring population and 9 percent of the total population.[75] By 1850, artisan and manufacturing establishments comprised 30 percent of the city's businesses.[76] Always Mexico's largest and wealthiest market, the city quickly moved to the forefront of early industry. By 1879 there were seven textile establishments in the Federal District and four in the surrounding municipios of Mexico State.[77] By 1910 there were dozens of textile enterprises in Mexico City, the Federal District, and the Valley of Mexico. These included a mix of large-, small-, and medium-sized enterprises, as well as the large commercial establishments tightly linked to the very largest textile firms. Numerous small sewing workshops, *boneterías*, were also abundant. There were none of the country's very largest mills, however.

The shift from manufacturing to industrialization during the Porfiriato, the natural growth of the city, and late nineteenth-century modernization pushed working class and poor families away from traditional urban centers and toward new peripheries, *colonias obreras* and *colonias populares*.[78] Within these neighborhoods, there developed the city's self-conscious working-class culture, which was later portrayed in film by artists such as Tin Tan and Cantinflas. Nonetheless, Mexico City was always too large and too diverse to become working class in the ways that would define the mill towns like Atlixco and Orizaba.

A 1924 census of the city's industries listed twenty-one textile establishments in the Federal District, mostly cotton. In general, the larger mills were more advanced, hired mostly men, and paid higher wages. Furthermore, the higher paid *empleados* were almost always men and

TABLE 2.6
Some Textile Factories in Mexico City, 1879

Place	Name	Number of workers	Owners
Distrito Federal	El Águila	225	I.R. Cárdenas y Cía.
Distrito Federal	Mercado de Guerrero	360	R. Arena y Hermano
Distrito Federal	La Minerva	160	Suinaga Hnos.
Distrito Federal	La Magdalena	520	Pio Bermejillo
Distrito Federal	La Hormiga	400	Nicolás de Teresa
Distrito Federal	San Fernando	142	Manuel Ibáñez
Distrito Federal	La Fama	220	Ricardo Sáinz
Estado de México	Miraflores	430	J. H. Robertson y Cía.
Estado de México	La Colmena	625	Francisco Arzumendi
Estado de México	Río Hondo	170	Isidoro de la Torre
Estado de México	San Ildefonso	111	Hijos de F. de P. Portilla

SOURCE: Mario Trujillo Bolio, *Operarios Fabriles en el Valle de México* (Mexico, 1997), p. 30.

sometimes foreigners. A sampling of a few large, medium, and small mills will provide some sense of the factories in the city.

La Carolina, owned by E. Noriega y Compania, was the largest mill. In 1921, it employed a thousand workers, who enjoyed the highest average wage in the city, 5.52 pesos a day. The 26 employees, foreign men, earned twice the average wage of male workers, who in turn earned 60 percent more than the few women employed there. The mostly male labor force (803 men, 21 women, 194 boys, and 10 girls) was entirely Mexican. Half the men were literate while only a third of the women could read.[79]

San Antonio Abad, owned by the company of the same name, employed 25 empleados and 501 *obreros* in 1920. All of the empleados were male, 10 foreigners and 15 Mexicans. The workers were all Mexican, 391 men and 110 women. The empleados earned an average of 7.76 pesos a day while the workers averaged only 2.78 pesos per day.[80] Electricity powered the 12,720 spindles, 352 looms, and 4 stamping machines, which produced 32,650 kilograms of spun thread and 3,027,637 meters of cloth. A labor inspector noted that "the factory is large, clean, and has sufficient ventilation."[81] The same company also owned La Colmena and El Barrón, two large mills in neighboring Tlalnepantla in Mexico State.

There were also smaller factories, like La Azteca owned by Antonio Gabriel. In 1921, it employed 20 men, 5 women, 6 boys and 4 girls. Although the minimum wage was 1.25 pesos a day, the men earned a

TABLE 2.7

Mexico City Textile Industry, 1924

Name	Location	Place	Workers	Average wage, pesos
Moíses Farja y Hno.	ciudad	ciudad	25	1.51
Tamacoco	ciudad	ciudad	36	2.49
El Salvador	ciudad	ciudad	271	2.56
La Guadalupe	ciudad	ciudad	209	2.84
La Victoria	ciudad	ciudad	333	3.00
Kuri Primos	ciudad	ciudad	55	3.00
C. A. Belmar y Cia.	ciudad	ciudad	202	2.93
San Antonio Abad	ciudad	ciudad	509	3.43
La Perfeccionada	ciudad	ciudad	775	2.87
Hipólito Chambón	ciudad	ciudad	194	3.6
Pasamanería Francesa	ciudad	ciudad	64	2.91
La Carolina	ciudad	ciudad	1,048	5.52
Manuel Castillo	Tacubaya	DF	13	1.05
La Abeja, S. A.	Tizapán	DF	454	5.5
La Alpina	Tizapán	DF	464	3.02
La Fama Montañesa	Tizapán	DF	485	3.5
La Hormiga	Tizapán	DF	962	4.00
La Magdalena	Contreras	DF	900	2.27
Santa Teresa	Contreras	DF	401	2.26
El Rosario	ciudad	ciudad	44	2.86
La Autora	ciudad	ciudad	152	3.71
City average				3.23

SOURCE: SICT, DT, Lista de Industrias establecidas en el Distrito Federal, que rindieron datos para el Censo de 1921, Hilados y Tejidos, January 25, 1924, AGN,DT, Caja 279, Exp. 5.

maximum of 2.50 per day, and the women and children earned only 1.50. Seven of the men were married and 13 were single, though undoubtedly many of the latter lived in *unión libre*. Only one woman was married while the other four were single; they may have had children but did not necessarily live in union libre.[82] In Tlalnepantla, Tomás de la Torre owned San José Río Hondo, located on the river of the same name.[83] In 1912, its 151 men, 12 women, and 16 children utilized 103 looms, 1,600 *husos de pie*, and 1,488 *husos de trama* to produce 67,000 pieces of manta each year.[84] The smallest enterprise in the city, that of Manuel Castillo, employed 13 workers and paid the lowest average wage, 1.05 pesos a day.

In 1923, the eighteen active cotton textile factories in the city used 130,527 spindles and 3,985 looms, employed 7,777 workers, and generated 4,917,575 kilograms of cotton and cotton products.[85] They employed more men than women and paid them more, though all workers earned less than the empleados, many of whom were foreigners. Literacy was not universal, but a considerable number of men and women could read and write.

Puebla and Tlaxcala

Within the Sierra Madre Oriental lie the states of Puebla and Tlaxcala. The central part of both states is a single high mountain valley to the east of the valley of Mexico. To the north, the mountains lead to Hidalgo and Veracruz, to the east there is the giant Pico de Orizaba, then the descent to Orizaba and the port of Veracruz on the Gulf Coast. On the west are the two large volcanos, Popocatepetl and Iztacihuatl, then the valley of Mexico. To the south is an extensive Sierra leading to Oaxaca. With independence, the valley formed the single state of Puebla; Tlaxcala achieved statehood in the 1850s. In 1909 Enock noted that "The valley of Puebla draws its varied sources of life largely from the Atoyac river, whose hydrographic basin forms a fertile region probably superior to any in the Republic. Level tracts of land and undulating valleys are irrigated freely from this river, giving huge crops of cereals, and numerous mills producing textile fabrics are actuated by the water-power it affords."[86]

The capital of Puebla state is Puebla city. While Tenochtitlan/Mexico City had been the Aztec capital before the Spaniards, Puebla was a new city, Puebla de los Angeles, established by the invaders in 1531. On the road between the viceregal capital in Mexico and the colony's main port in Veracruz, it quickly became the territory's second city. Like Mexico City, colonial Puebla was a textile and artisan center, the city's workshops producing not only cotton but also silk and wool. Mexico's first industrialists established cotton textile factories there, taking advantage of the Río Atoyac and a local labor force already familiar with the manufacture of cloth.[87] Locals considered it "a great industrial center, both manufacturing and industrial, an inexhaustable source of work and welfare for the proletariat, and a vein of wealth for the community, where the worker sector is very large because the complexity of work in the factories and workshops requires employing muscular force and machines in order to produce a variety of articles."[88] At the time of revolution it contained about 10 percent of the state's population, 96,121 inhabitants in 1910, 95,535 in 1921, and 114,793 in 1930.[89]

Like Mexico City, Puebla had a mixture of small- and medium-sized mills. In 1922, a Labor Department report listed sixteen in the state cap-

TABLE 2.8

Puebla, Registered Textile Factories, 1922

Factory	Location	Capital (pesos)	Male	Female	Children
			Number of workers		
Guadalupe	Puebla, Pue.	100,000	140		27
San Ignacio	Puebla, Pue.	50,000	112		68
La Providencia	Puebla, Pue.		197		17
El Volcan	Puebla, Pue.		220		27
San Jose	Puebla, Pue.	30,000	22	2	4
La Corona, S.A.	Puebla, Pue.	600,000	70	334	31
La Perla	Puebla, Pue.	50,000	54		20
San Alfonso	Puebla, Pue.	300,000	142		
La Economía	Puebla, Pue.	300,000	125		
Santa Ana	Puebla, Pue.	494,205	148	2	36
El Mayorazgo	Puebla, Pue.	4,000,000	649		70
El Esperanza	Puebla, Pue.	4,000,000	194	3	54
S. Juan de Amandi	Puebla, Pue.	4,000,000	407		36
Santiago	Puebla, Pue.	494,205	327		68
La Poblana	Puebla, Pue.	220,000	253	2	59
La Violeta	Puebla, Pue.	120,371	84		7

SOURCE: Cuadro Sinóptico de las Fábricas de Hilados y Tejidos que Han Rendido Sus Datos Estadísticos Relativos al año de 1922, AGN, DT, Caja 405, Exp. 1.

ital, ranging from El Mayorazgo with 649 workers to San Jose, which only employed 22. The labor force was overwhelmingly male, with the exception of La Corona, which employed 334 women and only 70 men. La Corona, however, was a bonetería, which explains the female labor force.

There were a number of other mills located in the state, as well as in nearby Tlaxcala, which was one of the country's smallest states, with a 1910 population of 184,171, a number that fell to 178,570 in 1921 then grew to 205,458 in 1930.[90] The most important mill in Tlaxcala was La Trinidad, located in Santa Cruz, 155 kilometers from Mexico City but only 37 from Puebla city. Owned by Manuel M. Conde, the mill had 320 looms, 7,000 spindles, and 361 *operarios* in 1913.[91]

Atlixco

Atlixco, some 30 kilometers to the southwest of Puebla city, became one of the two most important mill towns in Mexico. It is the *cabecera* of the *municipio* of the same name. It sits at the base of the volcano, Popocatepetl, which rises 5,452 meters above sea level. Popocatepetl has been spewing smoke since 1347 and is the source of the waters flowing down the mountain; their sudden drop in altitude makes the location ideal for water turbines and textile factories.[92] The 1857 Constitution divided Puebla and other states into political districts, each of which contained various municipios. Before the revolution, Atlixco was the seat of a political district with its own *Jefe Politico*. The 1917 Constitution suppressed the districts and their bosses, replacing them with the *municipio libre* and an elected *ayuntamiento* and *presidente municipal*.[93] In 1900, the municipio counted 53,000 inhabitants, of whom 26,000 spoke Nahuatl.[94]

Atlixco developed a textile industry in the latter half of the nineteenth century, much like Puebla. The French families Maurer and Leblanc and the Spanish family of Angel Díaz Rubín owned extensive haciendas in the region, including San Mateo, El Volcan, and El Carmen.[95] José Antonio Serrano, Spanish, owned the hacienda of La Concepción, where he established the region's first cotton textile factory in the 1840s.[96] Another Spaniard, Manuel García Teruel, built a mill, La Carolina, in 1870.[97] In 1899, Florencio Noriega, Inocencio Sánchez, and Juan Bannister, also Spaniards, formed Noriega Sánchez y Companía, which transformed Noriega's mills in the San Agustín hacienda into a textile factory, Los Molinos. The widow of Francisco M. Conde came to own San Agustín, and San Agustín Los Molinos continued as a cotton mill. In 1897, José Romano, José Villar Romano, and Enrique Artasánchez bought El Volcan and in 1899 turned it into a textile operation.[98] These Spanish entrepreneurs turned the flour mills on the old haciendas into cotton textile operations.

In 1898, Lions Hermanos, who already owned stock in the Companía Industrial de Orizaba (CIDOSA), got a concession to use water from San Baltazar Atlimeyaya, a river that runs down the volcano. Two years later they built a mill, El León, some eight kilometers from town. In 1899, Luis Barroso Arias formed the Companía Industrial de Atlixco S.A. (CIASA), which bought lands from the Maurer family in the hacienda of San Diego Metepec. The company then built the enormous Metepec mill, one of the largest and most modern in Mexico.[99] These two giants, El León and Metepec, turned Atlixco into a center of the country's textile industry. Finally, in 1908, Angel Díaz Rubín turned his mill into a textile factory, El Carmen.

By the revolution, Atlixco was a small town with seven factories and little else. The children of Angel Díaz Rubín owned two factories, La Concepción (La Concha) and El Carmen. Sánchez Gavito y Cía owned

TABLE 2.9

Atlixco Mills, 1921

Mill	Owner	Workers	Looms	Spindles
San Agustín	Sánchez Gavito y Cía	300	309	3,244
El Carmen	Hijos de Angel Díaz Rubín	230	232	6,040
La Carolina	Viuda de Ramón Gavito	200	117	4,776
La Concepción	Hijos de Angel Díaz Rubín	320	336	8,616
El León	Cía Industrial Veracruzana	525	426	6,848
Metepec	Cía Industrial de Atlixco	1,100	1,574	34,083
El Volcan	E. Artasánchez y Cía	200	230	5,096

SOURCE: Benito Flores, April 8, 1924, AMA, Presidencia, 1924, num. H3.

San Agustín, and the widow of Ramón Gavito owned La Carolina. E. Artasánchez y Cía owned El Volcan. These mills were all in town. El Carmen was on the edge of the Río San Baltazar. The large Cía Industrial Veracruzana owned El León, eight kilometers northwest of town. The Cía Industrial de Atlixco owned Metepec, the largest mill in the state of Puebla, fifteen kilometers northwest.[100] Metepec built housing for its workers, and its compound was one of the centers of the workers' revolution. The workers of the other factories mostly lived, shopped, and fought in a very working-class town.

Orizaba

Just as Atlixco sits beneath Popocatepetl, Orizaba sits beneath the Pico de Orizaba, the highest mountain in Mexico. With flowing streams to provide power, the town also sits astride the road from Veracruz (133 kilometers), the country's principal port, to Mexico City (292 kilometers), the principal market. Textile workers in the 1930s said Orizaba was "as interesting as Manchester in England."[101] They added that it "is the most important textile center in our country and perhaps in the Americas because it has a great production of spinning and weaving, that for their quality, manufacture, and fine finishing, rival foreign production in similar factory centers."[102] Cocolapam was built in the 1830s and remained the country's largest cotton textile mill in the first half of the ninetenth century. In the 1890s, the two industrial giants CIDOSA and CIVSA took control of the region's mills while adding others. CIDOSA bought the local mills San Lorenzo and Cerritos as well as Cocolapam. It then constructed the largest and most modern textile mill in Latin America, Río Blanco, in 1892.[103] Only Santa Rosa remained independent, owned by CIVSA. CIDOSA also owned three hydroelectric plants, Río Grande, Cocolapam, and Boqueron.

In 1907 an American writer commented that,

Orizaba is a town of thirty-five thousand people and is a very beautiful and interesting place with its palm-shaded streets and low Moorish buildings. Its Alameda is a quaint, shady park with an abundance of flowers and blooming trees. Along the street the orange trees thrust their laden branches out into the highway over the low adobe walls. On the banks of the stream the washer-women beat their clothes to a snowy white upon the smooth round stones.[104]

Orizaba itself was the principal municipio of the four municipios that comprised the region, Orizaba, Camerino Z. Mendoza (antigua Villa de Santa Rosa), Nogales, and Río Blanco, with populations of 42,952, 10,200, 7,215, and 10,000, respectively, in the early 1930s. For textile workers, "it is one entity."[105] The nearby countryside produced coffee, sugarcane, tobacco, corn, and beans.[106]

As a group, the region's mills were the largest and most modern in Mexico. Río Blanco and Santa Rosa were two giants of the industry. The other three mills were larger than most Puebla or Mexico City mills. Like the Atlixco mills, they employed few women. Orizaba was for a time a wealthy mill town, a man's world, and a center of the workers' revolution, one that could draw on a few other industries located in the region.

There were other cotton textile plants in Mexico, though no industrial zones to rival Mexico City, Puebla/Tlaxcala, Atlixco, and Orizaba. There were mills in Guadalajara, the large El Hércules plant in Querétaro, and some scattered, mostly small mills in other states.

Working Cotton

By 1910, the manufacture of cotton fiber in Mexico was a mature, disciplined, and hierarchal enterprise. Most factories imported their machines from England. The machines brought a management-controlled labor process that required strict supervision. An expert on the English industry noted that,

Good management is indispensable to success. The sequence of processes down to the minutest details must be perfectly familiar to the manager, who has to be able to so co-ordinate the productive system that the greatest weight of yarn is got off within the hours of running, at the least cost, quality of staple, of course, being a dominating factor . . . The erection of a spinning mill is an expensive undertaking.[107]

TABLE 2.10

Orizaba Mills, 1923

Factory	Ownership	Workers	Male	Female	Children
Cerritos	CIDOSA	442	406	0	36
Cocolapam	CIDOSA	666	592	36	38
Mirafuentes	John Harrison	170	156	14	
Río Blanco	CIDOSA	2,759	2,610	12	137
San Lorenzo	CIDOSA	784	711	30	43
Santa Rosa	CIVSA	1,558	1,364	99	95

SOURCE: SICT, cuestionario de huelgas, various, January 1923, AGN, DT, Caja 558, Exp. 4; Mirafuentes data is for 1921 and from SICT, cuestionario sobre trabajo, AGN, DT, Caja 299, Exp. 1.

Although not all mills did all things, the core industry transformed cotton fiber into cloth. The same author described the industry in his country at the turn of the (twentieth) century. Bales of cotton had to be delivered to the factory, where they were inspected for quality. "The mixing together of different varieties and staples of cotton is essential in order to get the average quality of the cotton used."[108] Although the mixing could be done by hand, there were also machines for that work. Then the cotton was opened and cleaned. Carding was an important part of this process.

The object of the carding engine is (1) to remove all impurities either natural or foreign in the cotton which have escaped the preceding processes; (2) the extracting of all short, immatured, broken or nepped fibres, the retention of which would weaken or otherwise reduce the quality of the yarn; (3) to disentangle the confused mass of fibres and lay them approximately in parallel order; (4) to attenuate or draw out the heavy sheet of lap into a thin fleece or film and contract it into a ribbon of cotton or sliver, fitted for the next process.[109]

The spinning machines provided the fiber with the right amount of twist and fineness. Then the mule was brought into place. "The object of the Mule is to draw the rovings into the required fineness, to twist the yarn to give it strength, and to wind on a short paper tube or on the bare spindle as may be required."[110] The cotton fiber would then pass to the weaving shed for winding, warping, and weaving.

"We have now reached the final stage by which the cotton is converted into a woven fabric, that of weaving. It will be seen how the textile processes are conducted in sequence, with the greatest expedition and economy of labour. The machinery in the various departments is driven by the powerful modern engine. It may be taken that a condensing horizontal engine of 250 indicated horse power will be required to drive 1,000 looms."[111] Each weaver operated two to six looms, "the average being three to four . . . The weavers are controlled by an overlooker who is responsible for the work . . . The overlooker, or tackler, sees to loom repairs, to the looms being supplied with warps, correctly gaited, etc., and he has an interest in doing all he can to keep up the production and therefore also the earnings of the weavers."[112] Finally, the cloth could be bleached, printed, and dyed, and then sent off to market.

When Esteban de Antuñano introduced mechanized production in Mexico in the 1830s, he brought English machines and technicians. As a consequence, the Mexican industry did not vary much from the English; it used the same carders, spinners, mules, and looms. There were no Mexican machines for the Mexican textile industry. In the case of England, technology corresponded to English inventiveness and English capital/labor requirements. In Mexico, the technology corresponded to conditions in England rather than conditions in Mexico. This resulted in large and complex Mexican factories that nonetheless always lagged behind English productivity. The need to import technology brought levels of productivity and mechanization that corresponded to economies quite different than the Mexican. As a result, Mexican textile factories were large, complex, mechanical operations, although never as advanced as foreign competitors with their constant innovations.

For cotton textile workers in the fiber transformation factories—the English used the more descriptive *operative*—the primary task was to operate the machines. Mexican *operarios* labored at an industrial rhythm imposed by the machines. They had to become part of the machine. Such workers, recruited from preindustrial textile manufacturing or from the countryside, had to adjust to factory time, factory responsibility, factory hierarchy, and factory authority.

In 1910, there were 127 cotton textile factories in Mexico that employed 22,000 workers.[113] In 1927, there were 159 factories with 41,000 workers, which included 5 large factories in Orizaba and 7 large- and medium-sized factories in Atlixco. The state of Puebla had 56 factories, mostly in Puebla city and Atlixco, while neighboring Tlaxcala had 8 factories. Mexico City had 27 factories, along with 8 in the neighboring municipios of the state of Mexico. There were 9 cotton textile factories in Coahuila, 6 in Jalisco, and scattered cotton factories in other states, from Chihuahua in the north to Chiapas in the south. The workers in these factories had much in common, including working cotton, life in the mill, textile wages, and their experience of class. One of the Puebla owners said, "The worker is the muscle of the industry. In workships, he carried out the muscular work alone. In the primitive factory system, he did it aided by the machine. In the current stage of industrialization, he carries out his work aiding the machine."[114] The owners saw workers as muscular subordinates to their expensive and sophisticated machines.

Directly operating machines, operarios had opportunities to either make their work easier or to increase their salary because most skilled workers received piece-rate pay. In 1912, the administrator of El Léon complained to the Labor Department that his weavers, "in order to make more cloth and make more money," regularly loosened the screws on the looms in such a way as to produce a greater quantity of a lesser cloth, "with prejudice to the interests of the Industrialists."[115] Workers may have assisted machines, but they often did it to their own advantage.

Conclusion

From its founding in the 1830s to 1910, Mexico's cotton textile industry was in the forefront of industrial change in the country. It brought the first true factories, the first industrial proletariat, and the first factory social relations of production. As the industry grew, it concentrated in four principal textile zones: Mexico City, Puebla city, Atlixco, and Orizaba. Ownership changed from a few entrepreneurs with foreign experience to large ownership groups, in which the French and the Spanish dominated. By 1910, increasing mechanization, a shift to electrical power, and more advanced machines brought larger mills and increasingly impersonal labor relations.

The rapid growth of Mexico's first factory industry brought new social relations to the country, but these were social relations that combined elements of the new with the old. Many of the textile mills were built in old haciendas with their traditional flour mills, and the owners of the industry continued, or tried to continue, with the paternalism of social relations in the hacienda. As in the hacienda, there was a sharp divide between owners, who looked down on workers, and workers who often resented their patrón. On the other hand, the mills imported foreign technologies, they grew to quite a large size, and they demanded a regularity of work patterns often absent in the old haciendas. Inevitably, new and more impersonal relations gave rise to new labor problems that the old social relations could not handle.

Furthermore, since modern industry in Mexico required knowledge of foreign technology, and since foreigners played a powerful role in the large-scale commerce that was so important to the textile industry, foreign owners came to dominate the largest and most modern mills. When the revolution exploded, nationalist workers could easily express their anger not only against owners as owners, but also against owners as foreigners.

Two other factors complicated labor relations in the industry. First, the industry eventually developed different kinds and sizes of mills with varying ownership patterns, from the individual owner of small, backward and specialized mills, to large corporations with large and very modern mills. With different concerns and needs, it was difficult for the owners to confront workers as a unified group. Second, it required a lengthy learning period and great skill to master the machines, so that many workers, particularly those with skills, understood their centrality to the industry, which strengthened their confidence in confronting the owners when times got tough.

In the next chapter, we look more closely at "the muscle of the industry." It came as a great surprise to owners, but these "muscles" could also think. When it came time to think revolution, radical thoughts and armed muscles transformed the lives of owners and workers alike.

CHAPTER 3

The Layered Communities of Mexican Cotton Textile Workers

EVER SINCE MARX WROTE *Capital,* revolutionaries debated the degree to which class was identity and whether class identity was necessary for revolution. For Mexican cotton textile workers, there was no debate. Social class was their identity, without which there would have been no revolution.

The development of an industrial working class inevitably accompanied the development of industry. The growth of one was the growth of the other. What was not inevitable was how that class conceived itself. Recruited from a mix of preindustrial textile workers and new recruits from the countryside, cotton textile workers by 1910 had almost a century of experience as subordinates, both in the factories where they worked and in the communities where they lived. However, this was never a homogenous group, at work or at home, nor were their experiences unchanging over time. Instead, a rather long history of varied experiences came together to form ideas of class that, given the right mixture of conflict in the factory, violence in the countryside, and a collapsing state, expressed themselves in Latin America's first workers' revolution.

The determining experiences of cotton textile workers came from their workplaces and their increasingly working-class communities. These experiences not only created a sense of an oppressed social class, but also layered feelings of community: the family, the workplace, the neighborhood or mill town, and their social class. These layered communities of Mexican cotton textile workers were the intellectual and emotional foundations of the workers' revolution within the revolution.

Interlocked and layered communities were not unique to textile workers but infused much of Mexican society. What was specific to workers was the degree to which their experience at work shaped their participation in and understanding of their respective communities.

This chapter organizes these communities into three broad areas: social class, social relationships, and social ideas. Each interlocked with the others to form a world view that allowed industrial workers to jump at revolution when the possibility presented itself. Primary, of course, was their concept of social class, without which there would not have been a workers' revolution.

Class

It is striking the degree to which textile workers identified themselves as members of a single social class, that of workers. They understood however that class did not mean homogeneity but rather appeared as social layers within which there was a common condition. Eighty years of industrialization shaped a view of self as subordinate wage laborers who, like all other subordinate wage laborers, constituted an oppressed class of people. They felt solidarity with others of their class and anger at those who oppressed them. Nonetheless, their workplaces did not treat them as an undifferentiated mass but instead divided them into layers, white and blue collar, skilled and unskilled, men and women, adults and children. The quality of their tasks, their working conditions, and most important, their incomes were a function of these layers. As a consequence, they identified as workers on certain occasions and in more specific ways on others. In both cases, their work experiences shaped their identities and drove their revolution. In the large mills, the majority of the labor force consisted of male skilled workers with relatively high wages and great pride in their skills. This group fought for the dignity of their social class but also to conserve a society in which skills and knowledge determined social rank. They led the revolution as workers but also as skilled men. The others followed.

Social Class

To understand their idea of social class, we can start with a police report of a murder in a textile factory in April, 1928. By this time the unions had won the workers' revolution, and union leaders had begun to separate themselves from rank-and-file workers. As part of that

TABLE 3.1

Statements to Police Official Manuel Aubry

Orizaba, April 12, 1928

Name	Origin	Class status	Age	Marital status
Reynaldo Pantoja	Oaxaca	dead	31	*soltero*
Gilberto Pantoja	Oaxaca		23	*casado*
Angel Trujillo	Orizaba, Veracruz	*obrero*	40	*casado*
Eugenio González Martínez	Orizaba, Veracruz	*obrero*	27	
Francisco Rojas	Puebla, Puebla		31	*soltero*
Francisco Sánchez	Santa Rosa, Veracruz	*obrero*	27	*casado*
Marcos Hernández	Puebla, Puebla	*obrero*	34	*casado*
Cosme Rosete	Puebla, Puebla	*obrero*	36	*soltero*
David Lara Pardo	Orizaba, Veracruz	*obrero*	35	*soltero*
Moíses Ibarra González	Orizaba, Veracruz	*obrero*	40	*soltero*
Mauricio Tobón	San Juan Ixcaquitla, Puebla	*obrero*	28	*soltero*
Andres Gines	Tehuacán, Puebla	*obrero*	49	*soltero*
Ramón Flóres	Orizaba, Veracruz	*obrero*	34	*soltero*
Gregorio Castellanos	Nochistlán Oaxaca	*obrero*	36	*soltero*
Patricio Rodríguez	Tecamachalco, Puebla			
German Tejeda	Atlixco, Puebla	*obrero*	25	*soltero*
Silvestre Rodríguez	Tlanepaquilla, Veracruz	*obrero*	28	*soltero*
Roberto Celis	Orizaba, Veracruz	*obrero*	29	*soltero*

SOURCE: copia integra de la declaraciones rendidas en la oficina de la policía judicial, April 12, 1928, AMO, Caja 201, Exp. 1.

process, the Cocolapam union expelled Mauricio Tobón and other members of the Communist Party who worked in the mill, which cost them their jobs. Tobón ignored the union's order to leave because "it is not the union leaders who should be firing workers."[1] When he defiantly entered the mill, union leader Marcos Hernández ordered work in the factory halted, then attempted to physically throw him out of the building. A fight ensued between followers of Hernández and those of Tobón. It is unclear from statements to the police whether the powerful union leader really enjoyed majority support, but in the fight somebody stabbed Reynaldo Pantoja with a large knife. Pantoja pulled the knife out, walked a bloody 160 meters, then dropped dead.[2] Tobón retreated to his house. The unionists assaulted the house at 6:30 in the afternoon, shooting Tobón in the stomach. He died in the hospital on April 30, fired by the union after all.[3]

Investigating Pantoja's death, the police took declarations from witnesses, a random sample of Orizaba textile workers.

As part of the interrogation, the police ordered the witnesses to state their name, place of birth, age, social class, and marital status. To social class, these workers in the mill responded with *obrero* or worker. They and the police understood that as social class. It was not just that they worked at the factory, because they could have responded *operario,* operator, or *trabajador,* from *trabajar,* to work. Instead they answered obrero because as factory workers they believed they were members of a larger social class, that of workers. *Operario* and *trabajador* lacked the class meanings of *obrero,* those who labored for a wage because of necessity. It was a class term that everybody in Mexican society understood.

The use of obrero was ubiquitous in early twentieth-century Mexico. When the mill hands at La Aurrera in Mexico City wrote to the national union in 1914, they asked for help in order "to avoid difficulties between owners (*patrones*) and workers (*obreros*).[4] In 1915, the Cerritos mill hands in Orizaba spoke of the "wide liberties that the Revolution has returned to the working class (*clase obrera*)."[5] When an Orizaba union leader complained about leftists, it was because they "have offended all the Workers in general (*Obreros en general*)."[6] When the San Lorenzo workers formed a union in 1915, it was to "guard over the interests of the Workers of this factory" (*Obreros de esta fábrica*) in order to deal with "the difficulties that arise between industrialists and workers."[7] The low-paid movers in El Léon accused owners of "creating difficulties, true pretexts that put into poverty the *working class.*"[8]

Social class was a distinguishing characteristic of society in early twentieth-century Mexico. An American describing the country noted that "What might be said of one class would not apply to another. The differences of dress and customs alone make known the heterogeneousness of the population."[9] A British observer commented that "Mexican society falls into lines of marked class distinction. The rich and the educated stand in sharp juxtaposition to the great bulk of poor and uneducated, and the high silk hat and frock coat form a striking contrast to the half-naked and sandalled peon in the plazas and streets of the cities."[10]

Allowing for individual differences, there was a general feeling among most workers that they were a humble, low-status, and disadvantaged class. In their letters, they described themselves as "suffering children of labor,"[11] "humble Servants,"[12] "a humble worker,"[13] "the poor workers,"[14] "suffering but hard working people,"[15] "humble and sad proletarians," and "we who are weak."[16] In 1915, the president of the Cerritos union argued that "They consider us workers as things and not as people."[17] In 1917, the Santa Rosa union sent "our fraternal salute to our brothers in misery."[18] The head of the Labor Office in 1914 described "the most humble classes of society."[19]

To many, obrero also signified workers' class solidarity against an oppressive hierarchy. An Atlixco union leader declared that strikers were "our brothers and sisters of work, of misery and of suffering." He added that the town's employees "have always expressed a certain hierarchy towards us because they dress better, and they don't want to associate with the true and humble working people."[20] Workers put up with these insults because, as a labor activist argued in 1912, we "need to work in these factories."[21] Necessity created the working class.

Neither class nor conceptions of class are static. The process of revolution not only changed the social relations of factories, but aspects of class distinctions. Initially, of course, Madero did not foresee social class as a component of his political revolution. However, the workers' uprising described in Chapter 4 uncovered a *problema obrero*. In response, Madero, once in office, established a Labor Department in order to find a "solution to the worker problem."[22] The new Labor Department noted that "as a general rule, in all the factories, workers and owners always consider themselves enemies."[23] These enemies were the two sides that fought the workers' revolution.

An important change that came with the revolution was the introduction of *obrero* as a legal term. While the 1857 Constitution and subsequent law mostly ignored work and workers, during and after the revolution *obrero* became a legal term, and *obreros* the subject of a large body of law, as described in Chapter 6. This began with the 1912 textile contract that made widespread use of the term, as in "every worker will occupy the place that he is assigned to."[24] Subsequent state labor codes also made *obrero* a common legal term. The Campeche code stated that the "the industrial agreement is an agreement between the representatives of the unions or any other grouping of *workers*."[25]

If, before the revolution, workers believed they belonged to a social class, after the revolution law defined that class, strengthening its usage. In the 1917 Constitution, Article 123 protected everybody who worked, *"obreros, jornaleros, empleados, domesticos y artesanos."*[26] It was quite specific in defining conflict between "capital and labor"[27] and in protecting the rights of obreros, such as when it defined certain rights when "the owner . . . fires a worker."[28] The state labor codes that implemented Article 123 followed this practice, so that the 1928 Veracruz State Labor Code outlined "the obligations of workers."[29] From a ubiquitous social term, *worker* became a ubiquitous legal term.

The process of revolution and its success strengthened appeals to class. For example, when a union opposed work rule changes in 1926, the workers argued that they would not have "our union rights trampled or weakened by our enemies the Capitalists; who try to humiliate us like in the past."[30] From Guerrero, Epitacio López of the Confederación Regional Obrera Mexicana (CROM) wrote that "the *compañeros* . . . live in complete slavery for a miserable salary . . . where is the necessary energy to avoid the attacks of Capitalism, our eternal enemy . . . today the bourgeoisie tries to terrorize our brothers, victims."[31] When Guillermo Palacios, president of the Comisión Especial del Salario Mínimo of Orizaba in 1933, filed his report, he spoke of "the class struggle [between] a proletarian class and the owning class of this place."[32] A tool of the workers' revolution, appeals to class became enshrined in Mexican society in the years following.

If textile workers saw themselves as oppressed, subordinate wage laborers, that is what eighty years of industrialization had taught them. But that was not all they learned in the factories. The mills were places of highly differentiated labor, in part based on skills and knowledge. Mill hands saw this and came to believe in it, so that they were not levelers. There is little evidence that they sought absolute social equality. Instead, many expressed an admiration for the knowledge and formal education that most workers lacked. In 1916, Río Blanco workers, concerned about low wages, circulated the idea of a Comité de Ajuste, with "an operator from each craft, in agreement with the head supervisor." Luis Osorio, prominent union member, expressed a fear that many workers lacked the necessary technical expertise to serve on the committee, suggesting that the company employ engineers in this capacity. Osorio explained that "we have not studied engineering . . . in that case, if we had, we would not have the jobs we have but better ones with better pay."[33] The skilled operatives of the textile mills recognized distinctions based on education.

Empleados

Workers approved of some distinctions based on education but not all. They disliked the differences between white collar (empleados) and blue collar (obreros), without doubt because of the deep social rift between the two. The former were few, tied to management, relatively well paid, almost always male, and virtually all literate. The latter were many, disdained by management, poorly paid in comparison, male and female, and their uncertain literacy usually not important to the factory. The empleados carried out managerial and office tasks and, at the beginning of revolution, considered themselves superior to workers. They earned a monthly salary, whereas the workers earned either a daily rate or piece rate.

The data from a few mills suggests the relative proportions of empleados and obreros. La Carolina in 1920 had 22 employees and 1,300 workers. All of the employees were foreign and male and earned twice the average wage of the workers.[34] La Abeja employed 27 empleados and 486 obreros. All the workers were Mexican while 6 employees were foreign.[35] The mill employed 309 female and 177 male workers, but no female employees. The male employees earned an average of 6.98 pesos a day, compared to 2.00 pesos a day for the male workers and 1.75 pesos a day for the women.[36] In La Pasamanería Francesa, the factory adminstrator was French and the director, Spanish, but the 213 workers—131 women and 82 men—were all Mexican. Of the ten employees, there were seven foreigners, among them, two women.[37] San Lorenzo listed eight empleados in 1922. One was the factory administrator, one was the cashier, and the other six were *oficinistas*.[38]

Before the revolution, workers and employees led separate lives. The empleados carried out the mandates of administration, so that workers saw them as a class enemy. During the 1911 general strike, empleados of the Santiago mill enforced the orders to prevent workers from entering the mill with their hats and *zarapes*, so the mill workers added to their strike demand that management fire all the empleados.[39]

The victory of the workers' revolution radically altered the relationship between empleados and obreros because it imposed unionization on employees. Their unions then became part of larger regional and national organizations that subordinated employees to unions rather than to administration. Empleados eventually joined the proletarian culture of postrevolutionary Mexico, using their literacy to rise through the union rather than management ranks. They often became allies rather than enemies of blue-collar workers, as we shall see in Chapter 9. Skill differences acquired a different meaning under widespread unionization.

Skilled and Unskilled

In a country where skilled jobs were scarce, the textile industry offered good work at decent pay. The larger, more modern mills mostly

TABLE 3.2

*Skilled and Unskilled Workers
in Selected Mills, 1921, 1922, and 1923*

	Number of skilled workers	Number of unskilled workers
La Carolina	1,110	86
Cocolapam	689	150
El Cisne	3	36
La Fama (N.L.)	60	28
Metepec	820	338
Miraflores	156	88
San Lorenzo	615	165
La Trinidad (DF)	57	86
La Union	21	13

SOURCE: Cuestionarios, AGN, DT, Caja 405, Exp. 1. For the number of unskilled workers in El Cisne, La Carolina, San Lorenzo, La Trinidad, and La
Unión, the occupations with daily wages below 2 pesos. For Cocolapam, the
same plus *peones,* who earned a bit more than 2 pesos. For La Fama, Cuestionario de salarios, July 1921, AGN, DT, Caja 275, Exp. 1, unskilled: 2 pesos and
below, skilled: above 2 pesos. For Metepec, Cuestionario de salarios, n.d., AGN,
DT, Caja 275, Exp. 1, skilled: 2 pesos and above, unskilled: below 2 pesos.

employed skilled workers with wages significantly above the minimum. It was a machine-driven industry in which detailed knowledge
of each specific machine was quite valuable. The larger factories usually employed a few supervisors, a large number of skilled operators,
and a smaller number of unskilled laborers. The chart above lists the
distribution of skilled and unskilled workers in selected mills.

Even among the skilled workers there were distinctions. At the top
of the hierarchy were the supervisors, often called maestros. *Maestro,*
from the Latin *magister,* was a common word with multiple meanings,
ranging from teacher to expert to boss. It could convey respect and
warmth in an everyday greeting or the proper distance when dealing
with a stranger. Among the community of skilled workers of the factory, a maestro was often one whose knowledge and work had earned
the respect of others. The mill elevated these men to maestro because
of their technical expertise and also because they commanded respect,
which they used to direct workers. They were, as Gramsci would have
put it, organic intellectuals from the working class, often at the service
of ownership.

The large mills were divided into departments, each with a main
task. Normally each department had a supervisor, or maestro, and the
largest departments, usually weaving, could have more than one,

depending on the number of shifts. In 1922, La Carolina employed four maestros, two for dyeing, one for spinning, and one for carding (cleaning). Each earned 10 to 13 pesos a day, well above the pay of ordinary skilled workers. It is notable that three of the four maestros were foreigners, as were all fifteen of the mill's empleados. The mill also employed two foremen (*encargados*). Both were foreigners earning 10 pesos a day. Common workers, on the other hand, were overwhelmingly Mexican. Of the 1,196 skilled and unskilled workers, only five were foreign born.

Among the workers, the vast majority were skilled, at least in the larger, more modern mills like Metepec and La Carolina. In Metepec, skilled workers outnumbered the unskilled more than two to one, 820 versus 338. Of the skilled workers, half were weavers, 419 men. These men earned an average maximum wage of 3.00 pesos a day, considerably higher than unskilled wages. For example, unskilled sweepers earned 1.11 pesos per day. In La Carolina, there were 1,120 skilled workers and only 86 unskilled. Among the skilled workers, the 450 weavers earned 2.50 pesos a day.[40] Cotton textiles was thus an industry of skilled operatives who took pride in their knowledge and their work and who often thought of themselves as artisans. A 1916 union meeting "gathered together all the artisans who comprise the personnel of workshops of the Río Blanco factory."[41]

The smaller, more specialized mills and knitwear plants often employed a greater proportion of unskilled laborers and women and usually paid lower wages. El Cisne was a small factory in Mexico City that manufactured thread. It employed sixteen men, twenty women, and a few boys and girls. Only four workers earned more than 2 pesos, and the others received unskilled wages. Whereas in the bigger mills the largest single job category was that of weaver, El Cisne listed its workers as diverse, meaning an assortment of unskilled or marginally skilled laborers.

Skill of course refers to the job offered by the mill, not necessarily the ability of workers. In April 1920, the Río Blanco mill rejected the petition of young movers for higher wages. The doffers (*mudadores*) had argued that their work was equivalent to that of folders. The factory responded that folding required "cleaner and more intelligent activity, demanding greater attention." It added that those who desired a higher wage could learn spinning, a skilled position with a higher wage.[42]

Workers themselves were conscious and proud of their skills. During the early revolution, weavers demanded that they not have to carry heavy items around the factory, which they said should be left to unskilled laborers. The owners ceded this in Article 19 of the 1912 agreement, which declared that heavy items would be transported "by peons or workers dedicated to this kind of labor."[43] Despite such feelings, there is little evidence of conflict between skilled and unskilled workers. To a large degree, unskilled laborers sought skills in order to better their lot, and both skilled and unskilled pursued educational goals consistent with their view of a natural hierarchy of knowledge. Additionally, many skilled workers began their careers as children doing unskilled tasks in the factory.

If there was little conflict between skilled and unskilled workers, there was no greater source of conflict than the relationship between maestros and common workers. In 1912, the head of the Labor Department declared that "most complaints are not against factory administrators but against the *maestros* of the workshops."[44] Most foremen had been skilled workers who had risen through the ranks yet had little difficulty shifting allegiances from workers to owners. When the maestro became the physical representation of the hierarchy and authority of the factory, conflict with workers became inevitable. As we shall observe in Chapter 5, workers could not assault authority without attacking foremen, and they could not attack foremen without assaulting the authority of capital.

Men and Women

Not surprisingly, early twentieth-century Mexico made a sharp distinction between the genders. During the Pantoja murder investigation, the police interviewed those on the shop floor closest to the murder, seventeen men and no women. As this suggests, very few women worked in Cocolapam or the other large mills. Cotton textiles was a man's world. In 1922, Cocolapam employed 754 men and 35 women. In 1927, the entire industry employed 32,112 men and 6,255 women. The gender division of labor was even sharper than these numbers suggest. Men occupied virtually all of the skilled positions. In the largest mills, they also comprised most of the unskilled labor. Mexico City's largest mill, La Carolina, employed 1,270 men and 30 women.[45] The much smaller San Antonio Abad factory occupied 412 men and 94 women.[46] The country's largest and most modern mills were in Orizaba. Río Blanco had 2,610 men, 12 women, and 137 children, and 1,364 men and 99 women worked in Santa Rosa.[47] Atlixco

TABLE 3.3

Mexican Cotton Textile Industry, 1927

Gender Composition of the Labor Force

Region/State	Active factories	Men	Women	Children	Total
Center	102	20,284	2,795	1,970	25,049
DF	24	4,622	1,471	381	6,474
Mexico	7	1,489	167	88	1,739
Puebla	53	9,929	365	1,269	11,563
Tlaxcala	7	1,579	—	129	1,708
North	18	2,980	1,355	217	4,552
Gulf	11	6,533	286	469	7,288
Veracruz	11	6,533	286	469	7,288
Pacific	13	2,315	1,819	215	4,349
Mexico	144	32,112	6,255	2,871	41,238

SOURCE: SICT, *Monografía sobre el Estado Actual de la Industría en México* (México, 1929), 39.

too was a man's world. In 1923, El Volcán employed 190 men and no women; La Concepción employed 355 men and no women; El Carmen employed 250 men and no women; and Metepec, the largest mill in the region, employed 1,399 men, 126 boys, and no women.[48] In Tlaxcala, La Trinidad employed 500 men, 50 children, and no women.[49]

The large factories, complex machines, and rigorous work had little to do with the maleness of the industry in Mexico. International scholars have noted that in much of the world, "The textile industry has generally been viewed as an industry suitable for the employment of female workers. In part this has been due to the tradition of textile production (particularly spinning) as 'women's work' in the pre-industrial household economy, and later its classification as a light (rather than heavy) manufacturing industry ... and therefore seen as suitable for the 'nimble hands' of female workers."[50] Most Mexican factories bought their machines from the British. When R. J. Peake wrote a manual of the British industry in 1910, he filled his book with photographs of female textile operatives operating English machines. Obviously women could work in textile factories.[51] In other parts of the newly industrializing world, "well over half the early labor force in cotton production, from New England to Belgium, was provided by women."[52] Mexico, however, was an exception for reasons neither industrial nor technical. It was the gendered construction of work rather than machinery that determined a male labor force.

Early twentieth-century Mexico defined men and women through their relationship with the family. Mexicans expected women to keep the family together, to bear and raise children, to teach them morality, to keep the home, to cook, to clean, and to look inward. L. Folsa, a female Labor Department inspector, said that schools needed to teach young girls to be "humble and consequently, obedient and thrifty." She counseled men to "help the wife and not ignore nor humiliate her."[53] This intelligent, progressive and educated woman defined women's role as subservient. On the other side, Mexican society expected men to work and to confront the external word. When Mexico industrialized and the factory replaced the countryside, everybody assumed that the labor force would be male. There was little public discussion of what society expected and demanded of women because most people understood and agreed. When Margarita Flores sent a complaint to Martín Torres, municipal president of Orizaba, she identified herself as "Margarita Flores, adult, single, occupied in the tasks of her gender."[54] She did not need to explain the "tasks of her gender" because *las labores de su sexo* were commonly understood.[55] Women and men shared these views. Born in 1900, Rosario Molina Parra moved to Atlixco, where her family worked in the textile industry. She commented that "almost all the women we were very foolish. In the first place, we weren't civilized [lacking] school; we didn't know how to earn a living in anything other than household tasks."[56]

In 1914, the Labor Department asked a group of its women investigators to present their ideas on education for women. It is remarkable how much these educated, professional, working women shared views of the role of women in society. Otilia Zumaya made it clear that education for working women had to serve the primary goal of becoming "a good mother and housewife." If poor, education would help her because "she will know how to be a worker but one intelligent in making clothes, hats, and embroidery."[57] Adela G. Viuda de Ysassi agreed with the others that the goal was to "train good mothers for the family."[58] As Ma. Ana E. Viuda de Morelos noted, to be a "good mother and good housewives."[59] If "fortune smiles on her, she could become the thrifty, good, and loving wife in a tranquil and happy home."[60] In short, they believed that women needed to be mothers and housewives first and only work outside the home if poverty befell them. If a woman had to work, Rosario Rodríguez argued that she should maintain the qualities of a good wife, docile and subordinate. "She will make of the Mexican working woman, a model, docile and selfless, with some degree of goodness and virtue, with which society will have gained much."[61]

Of course, Mexican social reality was often quite different from idealized reality. In theory, men supported women, their wives, mothers, and daughters. In practice, many women, both married and unmarried, worked outside the home. John Lear found that in Mexico City, "the majority of poor and working-class women could rarely afford to withdraw permanently from the labor market."[62] Some women suffered from husbands with insufficient incomes. Others lacked husbands but had families to support, children and sometimes parents. Some were unmarried, sometimes young, many times not, and lived with their families, contributing an income. The social reality of Mexico was that many women worked, and many women lived without men. It was often a hard life, quite distinct from the housewife idealized by society.

When they found a job in textiles, women discovered that most workplaces treated them differently than men and not better. An argument for that, as expressed by a male federal labor inspector, was that "women can never have the same energy in order to yield the same work as a man."[63] Another was the common belief among male owners that women, regardless of age, were girls who lacked a man's capacity for rational thought. William Jenkins labeled all three hundred women in his La Corona factory, "*niñas*" regardless of their age.[64] This set of beliefs—that men financially cared for the family, that women should be at home taking care of children, and that women were physically and mentally weak—condemned women to a marginal role in Mexico's most advanced industry, cotton textiles. Skilled work in cotton textiles was a good job, but it was usually denied to women.

The Mexican labor market was thus gender stratified by both industry and occupation. This segregation resulted in women working mostly in low-level occupations in the more backward factories. The low-paying knitwear plants, related to the traditional woman's occupation of making clothes, often employed a majority female labor force. The smaller, less modern mills too employed women. On the other hand, the largest and most modern mills often employed no women, and when they did, it was usually for the relatively unskilled occupations. Women could not aspire to the best jobs in Mexico's best industry.

La Perfeccionada was one of the larger knitwear plants in Mexico City. Like many others, it preferred women (528) to men (57). Of the fifteen departments, seven employed only women, including the large drawing and sewing departments. Two small departments, mechanics and dyeing, employed only men. Thus nine departments were fully gender segregated.[65] Of the six departments that employed men and women, the occupations were segregated. Weaving largely employed women (93). Five men were all mechanics. Working with the spring machinery employed eight men—two supervisors, a guard, three machinists, and two others—but no women. In shirt packing, four men worked the press but there were no women. The only two women in the warehouse were cutters, a job that did not employ men. In hosiery finishing, the only man worked a machine for which women were not employed. So even though La Perfecionada employed both men and women, it gender segregated every single job or department.[66]

Gender segregation of jobs was an important element in determining that men earned more than women. Table 3.4 on the opposite page indicates male-female wage differentials. It does not include the largest Orizaba and Atlixco mills, which paid the highest wages and employed no women at all, so that actual differences throughout the industry were greater than seen in the chart. Mexicans expected women to earn less than men. Manuel Díaz, federal labor inspector, visited the small cotton mill La América in Léon in 1913, stating that "the first thing to note is that 90% of the workers are female, for which reason they accept earning little."[67]

The mills were a man's world, but since about 20 percent of the labor force in cotton textiles were women, three questions emerge about maleness and revolution in Mexico. First, did maleness shape the workers' revolution in Mexico's textile industry? Second, did men and women behave differently during the workers' revolution? Third, did the revolution equalize male and female working conditions?

The last two questions will be addressed in the following chapters. Here it is important to note that many Mexican men grew up in a culture that stressed honor and dignity. They learned that being male meant defending oneself and one's family with physical courage and the willingness to use and confront violence. The men learned a kind of violence that became useful once the revolution began. Like the surrounding revolution in the countryside, the revolution in the factory was also violent. That is why Pantoja's enemies stabbed him to death, a frequent occurrence in Mexican factories during the years of revolution. There is little evidence that women workers stabbed enemies with the frequency that men did, though women workers actively joined the strikes and did engage in some violence. Thus the owners saw men as more difficult than women and women as more docile than men. Jenkins told the Labor Department in 1912 that "I wish to change the men for women, because you know better than anybody the difficulties we

TABLE 3.4

Male and Female Average Daily Wage In Selected Mills, 1922

Mill	Location	Number			Average wage		
		Men	Women	Children	Men	Women	Children
La Corona, S.A.	Puebla	70	334	31	2.25	1.28	0.80
La Fama Montañesa	D.F.	298	95		1.25	1.35	
El Hércules	Querétaro	698	240	139	2.47	0.97	1.14
Pasamanería Francesa S.A.	D.F.	73	132	12	2.80	1.50	1.30
Río Blanco	Orizaba	2,563	47	136	8.55	2.72	1.17
Río Grande	Guadalajara	784	389	191	2.50	1.33	0.65

SOURCE: Cuadro Sinóptico de las Fábricas de Hilados y Tejidos . . . 1922; for Río Blanco, DT, Sección de Estadística, AGN, DT, Caja 405, Exp. 1.

have had with cotton factory workers in this district."[68] To the owners, the labor problem was a male problem.

Children

When Dr. Ambrosio Vargas, "the only doctor in this town," visited the Santa Rosa mill, he attended "a young thirteen-year old named Daniel Gasca."[69] Vargas noted that the boy, who worked in stamping, injured himself when he "imprudently" came too close to one of the machines. He said that Daniel would recover from his wounds but expressed no surprise about an adolescent working in the factory. Many children worked in Mexican mills. As one mill hand later noted, "having grown up in this, I began to work at the age of 12."[70] In 1920, the four large Mexico City mills, La Magdalena, Santa Teresa, La Hormiga, and La Alpina, employed 241 children, almost 10 percent of their labor force. The average wage for the children ranged from 90 centavos to 1 peso, considerably below the average adult wage.[71] Although these mills only hired boys, those that normally employed women also hired girls. El Salvador for example employed 83 men, 5 boys, 130 women, and 24 girls.[72]

Many of the boys entered the factory at a young age to learn a skill that would allow them to ascend in rank. Reyes López, a native of Orizaba, began to work in Cocolapam when he was twelve. He stayed there for six years, left the region for two, then returned in 1917 to work as an unskilled folder before obtaining a promotion to weaver.[73]

Factories hired children for unskilled tasks that required little education, training, physical strength, or judgment and because they could pay them less. Many mills reserved certain positions for children. La

Esperanza, for example, reserved three of its twenty-four job categories for "boys," who earned the lowest wages in the factory. In that factory as in others, the men earned more than the women, who in turn earned more than the boys.[74] Mills generally stratified occupations by skill, gender, and age, paying salaries accordingly, or as Santa Rosa said, "the wage he earns is that which corresponds to him, as a boy."[75]

Thus, in the mills and on the shop floor, workers learned class as subordinate wage labor, but they also learned the stratifications and layers of class, which they rarely challenged. Their work shaped their conception of class. They were one in opposition to capital, but many in the different tasks and occupations of the industry. Their revolution allowed them to challenge capital more efficiently than the differences within their ranks.

Relationships

Despite a rather long workday, most textile workers did not live to work but rather worked to live, to enjoy family and community. These provided a sense of identity at least as strong as social class, and for many, stronger. While most observers commented on the strength of family ties, ties to the physical communities of region, town, and neighborhood were also deep. With regard to region, migration to textile zones often meant a difference between place of origin and place of work. Nonetheless, whether they lived in a textile city like Puebla or Mexico, a textile town like Atlixco or Orizaba, or a mill town like that of Metepec, family, neighbors, friends, enemies, and acquaintances gave social life a purpose, a context, and another set of layered communities that shaped workers' visions of the world.

Origins

Most Mexicans identified strongly with their region of origin, though these regions varied greatly in a country as large, diverse, and changing as late nineteenth- and early twentieth-century Mexico. From a very Indian south to a modernizing and thoroughly Spanish-speaking north, what seemed to characterize the country was that it was rural and agricultural. In 1910, most Mexicans lived in the countryside (71.3 percent) and worked in agriculture, which employed four times more people than manufacturing, mining, and construction together. Employment in manufacturing was dynamic and grew steadily through the late nineteenth and early twentieth centuries: 554,555 people in 1895, 613,913 in 1910, 534,428 in 1921, and 696,161 in 1930.[76] Even so, industry was an urban island in a rural sea.

Because industrial expansion took place within a rural milieu, it is natural to think that most textile workers came from a rural background and represented a group whose migration implied shifting regional loyalties. This was generally not the case, however. By the early twentieth century, most textile workers came from urban families, often with a history in the textile industry. In part this was because cotton textiles was an old activity in Mexico, predating industrialization. It was also because industrialists built the new factories in the old, preindustrial textile zones, like Puebla and Mexico City. Finally, the newer Atlixco and Orizaba mills recruited workers from the older textile zones. In her study of Atlixco, Leticia Gamboa found that most mill hands came from Puebla, Tlaxcala, Mexico State, Veracruz, or Mexico City, all textile centers. Less than 5 percent were from other states. Fully 31 percent came from the city of Puebla, and another 10 percent were native to Atlixco.[77] Looking at the El León mill in 1905, Gamboa found that half the mill hands came from the city of Puebla. The other half came from municipios in Puebla and Tlaxcala as well as from Mexico City and Orizaba.[78]

John Lear's research on Mexico City also suggests urban and working-class backgrounds for textile workers. The city's population jumped 28 percent between 1900 and 1910, with most of the new immigrants from central Mexico.[79] "How much of this population came from rural or urban areas is not evident from census data for this period, but all the major feeder states except Michoacán were relatively urbanized." Lear concluded that the pre-1910 migration to Mexico City was more urban than the post-1945 migration, which is consistent with the data from Atlixco and Orizaba.[80] Without doubt, urban and textile origins contributed to the strong sense of class that we noted earlier among the industry's workers.

Nonetheless, the late nineteenth-century expansion of the textile industry did attract some new workers from rural areas, though mostly those adjacent to the textile zones. Bernardo García found that most Orizaba workers came from the traditional textile centers, but there was a secondary migration from poor agricultural regions, particularly neighboring Oaxaca.[81] In one small Puebla mill, the federal labor inspector found it difficult to draw the line between rural and urban. He reported that the owners complained about "a lack of practice among

TABLE 3.5

Mexico

Total Population and Economically Active Population

1895–1930

	1895	1910	1921	1930
Total Population	12,632,427	15,160,369	14,334,780	16,552,722
Male		7,504,471		
Female		7,655,898		
Economically Active Population		5,337,889		
Male		4,802,734		
Female		778,559		
EAP by Sector				
Primary activities		3,584,191		
Secondary activities		803,262		
Tertiary activities		884,589		
Workers in manufacturing	554,555	613,913	534,428	696,161

SOURCE: INEGI, *Estadísticas Históricas de México*, 1994, 42.

workers, who are mostly from the countryside . . . although most have previous experience working in this factory."[82] Alan Knight has described this industrial complex surrounded by the overwhelming countryside as "an industrial *rus in urbe*."[83] In late 1921, federal labor inspector Roberto Saviñón toured the Puebla/Tlaxcala region, whose mill owners were laying off many workers because of the country's difficult economic situation. Saviñón commented that "Most of the fired workers are campesinos who have lands that they work."[84] While most Mexican textile workers came from and lived in working-class neighborhoods and communities in Mexico City, Puebla, Atlixco, Orizaba, and other towns, many retained close ties to the Mexican countryside.

Home and Community

Working-class families generally lived in modest residences, but home—*hogar*—had great meaning for them. For textile workers, their physical homes—*casas*—and communities were largely a function of the size and location of the mills. Large factories located outside urban centers usually provided housing for workers. In this, the large mills resembled the old haciendas that had given rise to them, with housing on the grounds for those who worked there. The compounds, like those

TABLE 3.6

Population of Textile States, 1910

Textile state	Population	Textile state	Population
Mexico City	720,753	Tlaxcala	184,171
Mexico State	898,510	Veracruz	1,131,859
Puebla	1,101,600		

SOURCE: INEGI, *Estadísticas Históricas de México*, 1994, 15–25.

at Metepec or Río Blanco, were villages unto themselves, with hundreds of textile families and thousands of people. During the revolution, the ease of communication inside the compound made them cauldrons of social revolt.

In the cities, most factory owners did not provide housing. Workers usually lived in working-class neighborhoods, side by side with workers of other industries, artisans, and families engaged in petty commerce. Both Mexico City and Puebla had numerous working-class neighborhoods, some located near the large mills on the outskirts of the cities, others near the downtown area where many small shops were located. The owners of mills inside the towns of Atlixco and Orizaba also did not provide housing. These towns were small and dominated by the large textile factories, so that families lived in working-class neighborhoods with many textile families as neighbors. The nature of housing determined a physically concentrated textile proletariat.

The houses themselves varied between the cities and the mill towns. In Metepec, the factory owners built a uniform set of two-room adobe houses laid out in rectangular rows. There was a large water tank in the center of the mill town for washing clothes and for sanitary facilities. The mill owners required workers and their families to carry an identification card or they would be arrested by the Rurales stationed inside factory grounds.[85] Río Blanco mill owners provided two-room houses of dry masonry. San Lorenzo factory owners also built dry-masonry housing with two or three rooms.[86] When federal Labor Inspector Manuel Ortega Elorza visited mills in Puebla and Tlaxcala, he reported good, clean housing. Most mills, he commented, had schools. Some provided theaters "where they moralize the workers as well as entertain them."[87] He added that El León's houses were "good given the very cheap rent."[88]

The companies charged the workers rent and also used their ownership to control the community. During the 1912 textile convention, Lic.

Martínez Carrillo spoke for the companies when he argued that the mills did not rent the houses simply for rental income but instead "these same houses have been built to service the factory; and we can say theoretically the worker can occupy the house while he is at the service of the factory."[89] The mill owners restricted visitation rights and always threatened to evict striking workers, both a source of irritation for workers families. Control over the houses became an explosive issue during the 1911 general strike and remained so throughout the revolution.[90]

Unlike the uniform housing in the mills, conditions varied in the cities, though most working-class housing was quite humble. In Mexico City, one working-class family of four rented a space three meters by two and a half meters for 6 pesos a month. The owner painted the room twice in a twenty-year period. A neighboring *vecindad* with a leaky roof also rented for 6 pesos. Housing six people, it measured four meters by three and a half meters.[91] As one writer said, "Whoever visits the houses of poor people, the accessories in which workers live, will be convinced that they lead a miserable life . . . In one small, dark, narrow and poorly ventilated room, many people of different ages and sexes squeeze together."[92]

Thus housing contributed to the workers' view that theirs was a humble social class. In spite of this, many families took pride in home and community. A 1932 report proclaimed that "Orizaba is a beautiful city that by nature, by the variety of climatological conditions, and by geography, is quite unique in Mexico, still maintaining the romanticism given to it by the Spaniards during the Colonial period."[93]

Family

The strongest community of all of course was family. This was as true for textile workers as for other Mexicans. As in most societies at that time descended from Rome, family identity was often prior to, stronger than, and inseparable from individual identity. Federal labor inspector Adela Vda. de Ysassi warned against "blemishing the family name, which because it is of your ancestors . . . ought to be venerated, forming an altar inside your breast.[94] In accordance, textile workers commonly framed their arguments as family necessity. Río Blanco workers launched a labor action in 1908 because of "our domestic misery, day after day life at home is more difficult."[95] In 1927, E. J. Hernández asked for a raise because of a salary, "not sufficient for me to take care of the needs of my family."[96] Being "submerged in the most critical hopelessness with my family" made Brígido Florez request work from the union in 1926.[97] Leopoldo Lezama wrote the union to ask for a job "for my brother Ramón so that he can get work and help my father and even more, my mother." He added that his father requested the solicitation and "there was nothing for me to do but obey."[98]

Even when constrained by a thirty-minute lunch period, most mill hands ate lunch with the family.[99] Mill hand Federico Modrid commented that "Sundays are days dedicated to rest, to clean up, to amuse oneself . . . that [if one worked on Sunday] he would deprive himself of resting tranquilly at home."[100] When companies threatened what workers viewed as family time, they protested, and such protests became a driving force in the workers' movement.

Family and formal marriage were only indirectly related. After the Liberals took power in the 1850s, there were four ways to be married in Mexico. For purposes of property and inheritance, only the state could perform and ratify a marriage. The state did not recognize Church marriages. In overwhelmingly Catholic, albeit partially Jacobin Mexico, the Catholic Church did not recognize state ceremonies and claimed for itself the sole authority to legitimize a couple's union. Some couples married in a state ceremony with a local judge, others in a Church ceremony with a local priest, others both, and still others, neither, living together in unión libre, or free union.

The police declarations in the Pantoja murder case are ambiguous. When the witnesses declared marital status, four of the eighteen adult men, ages twenty-three to forty-nine, replied married, while the others noted *soltero* or single. Without more data, however, it is difficult to know their private relationships. It is probably the case that some who claimed single status actually lived with a partner in some sort of permanent relationship. We know that with the exception of migrant laborers, the homeless, and a few others, most Mexicans did not live alone, but instead with family or relatives. Other factory lists resemble the Pantoja reports. Lists from San Antonio Abad in Mexico City and San Juan Amatlán in Puebla show a large number of apparently single workers. At La Fama Montañesa in Mexico City, the list showed 236 married men, 70 married women, 84 single men, and 46 single women. Although it is hard to know the true marital state of those who claimed single status, it is very likely that many who claimed to be solteros did so because they lived in unión libre. As a group, there were 482 children.[101]

TABLE 3.7

Family Size Among Workers in Two Mills

Mill	Workers without family	Families of two to five persons	Families with more than five persons
San Juan Amatlán	146	98	116
San Antonio Abad	406	210	24

SOURCE: For Amatlán, Francisco C. Conde to Srio de ICT, 24 de agosto de 1920, AGN, DT, Caja 299, Exp. 1; for San Antonio Abad, San Antonio Abad, Cuestionario sobre trabajo, 23 de agosto de 1920, AGN, DT, Caja 299, Exp. 1.

Perhaps less ambiguous is housing data, because if unmarried couples lived together, it would show up in these reports. A report on living conditions in two *viviendas* showed the number of people occupying one-room apartments. In eleven apartments there were couples in each, in nine there were four people in each, in three there were eight people in each, in eight there were six people in each, and in seven there were five in each. In these one-room apartments, only one person lived alone, while one apartment housed ten people. The apartments measured four meters by three meters and lacked private bathrooms, and the common bathroom was unclean and unwashed.[102] Though we may not always know marital status, the report confirms that very few Mexicans lived alone. Almost everybody lived with family, however it may have been constituted.

Just as the real lives of women differed from their idealized lives, real families often varied from the social ideal of husband, wife, and legitimate children, although there were numerous legally married husbands and wives living with their legitimate children. In urban Mexico, many men had affairs outside marriage. Although with greater difficulties, women also had affairs. Affairs often resulted in children, so that while some men lived with their legitimate wife and children, others abandoned them to live with the new girlfriend and children. When Facundo Pérez entered work in La Carolina in Atlixco, he lived with his uncle Aureliano Ramírez, "father of three, four or five children, with two or three women."[103] Facundo spoke with pride rather than shame of male virility. For women, however, such male virtues made for single mothers who often had to raise children, care for parents and siblings, and work full time in a textile mill for the lowest wage while suffering the shame of an extramarital affair. Whatever their feelings, many women were the unmarried heads of household with one or

TABLE 3.8

Number of Persons Living in One-Room Apartments
Mexico City Housing Inspection by J. de Beraza, June 1, 1920

People in each apartment	Number of apartments	People in each apartment	Number of apartments
1	1	6	8
2	11	7	1
3	8	8	0
4	9	9	0
5	7	10	1

SOURCE: J. de Beraza to Julio Poulat, 1 de junio de 1920, AGN, DT, Caja 223, Exp. 10.

more children to support. Often parents, aunts, uncles, brothers, or sisters lived with the family or with the unmarried mother. Real families were often quite different from the ideal but they were family nonetheless.

The families were often quite large by modern standards. When complaining about working conditions in the Orizaba mills, Enrique Hinojosa told the Minister of Interior Rafael Zubirán Capmany that the workers had four, six, or eight children to support, plus elderly parents, and "it is not possible to cover all the expenses." [104]

Among the expenses was food. In an era before fast foods, family meant food. Most Mexicans ate with the family. Factories did not provide food for workers, nor did Mexicans enjoy bringing foodstuffs to the factory. In the mill towns, workers went home to eat, or rather, the male workers returned home where their wives had a meal ready. Orizaba mill hands told Zubarán Capmany that before the revolution:

in order to eat our meals, they usually gave us a half hour in the morning, but in the majority of the factories, from the center of the departments where we worked to the gate to the street, there is a rather large space, in order to leave one took seven, eight, even ten minutes, as much for the amount of people, quite numerous, as for the distances, it is easy to understand that to get to our houses we needed a quarter of an hour, and for that reason, we had to return to work almost without eating. Now all of that has changed, because in each strike in the mills, much blood was spilt.[105]

This comment suggests that the gendered division of labor in the mills also characterized families. By and large, women cooked, cleaned, and raised the children. Men helped if they were willing to be labeled *mandilón* by their friends. A 1925 report on work hours noted that "The worker who must enter [the factory] at 7 am, needs the wife to get up at 5:30 at least, or at 6 in order that she can prepare his breakfast."[106] In 1912, Puebla strike leaders reported that the general strike in the textile industry broke out because:

in a single factory . . . the administrator or the manager ordered that workers without families could not eat in the houses of others who voluntarily agreed that their families provide them food, something absolutely indispensable, because without this help, workers without families would not have a place to eat, because in general the factories are not located in towns but some distance away, which impedes single workers from having a place to eat.[107]

Mexicans did not doubt that a woman "has her mission in the home."[108]

As we have seen, however, home for women who worked in textile mills often varied the social ideal. Most lived in one of three situations. Many stayed with their parents, working to help the family escape poverty. If these women chose to not marry, they would continue to live with their parents, and if the parents died, they would often live with the family of a brother or sister, still working to contribute to the family's income. Other women lived with men who earned small incomes, so that they worked to help sustain the new family. In such cases, they had primary care of the children and primary responsibility for cleaning and cooking, as well as their work outside the home. Finally, many women had children and or parents to support and lived without a husband, so they worked to support the families that they headed.

Whatever their family situation, most women textile workers were relegated to low-paying jobs. A Labor Office report described a woman who worked in the garment industry. She was illiterate and lived with her elderly father and niece. Her income with that of her father's, a fruit vendor, barely covered their modest expenses. She worked from eight in the morning to seven in the evening, six days a week. During her lunch hour, she had to buy tortillas and cook a meal for her family.[109] Another report on a seamstress noted that she supported her sister and three children with her 75 centavos a day income.[110] Another seamstress supported herself, her elderly mother, a sister, and a son on 48 pesos a month, of which 12 paid the rent for the small room where they lived.[111] As well as working in the factory, she cooked and did the laundry.

For men and women, whatever their specific family situation, family came first. In the mills, workers sought to get jobs for family members. In large part, their struggle to gain control over hiring was in order to facilitate jobs for family members. Family first also made shop-floor conflict a family matter. To a degree, the workers' revolution was a revolution of textile families.

Friends and Enemies

Personal relationships mediated between family and community. Though family was first, friends and enemies played important roles in social life. During revolution, the spread of violence meant that friends and enemies often determined life and death. Friendship was so important and so all pervasive, people signed their letters "remaining your friend" even when no actual relationship existed. [112] The opposite was also true, because honor demanded that no insult remain unanswered.

As we have seen, personal relationships influenced hiring, although the revolution changed the mechansim. Before the revolution, a mill hand might recommend a son, a nephew, a friend, or the son of a friend to a supervisor or the factory administrator. Sons followed their fathers and older brothers into the mills, and friends brought friends. After the revolution, workers recommended friends and family members to the union boss rather than factory administrator. A 1927 Metepec report showed that Manuel Pérez Pelaez "was hired on the recommendation of a friend."[113]

A 1956 study by Gabino Islas reported that the mill operated on the traditional system in place for a century. The workplace was a network of family and friends. The union hired the workers through personal relationships. Once in the mill, the apprentice was assigned to "a family member or friend of the family" for instruction.[114] As one union noted when it allowed a punished member to keep his job, he "will continue being everybody's friend."[115]

The Pantoja investigation revealed a fight between two antagonistic groups of friends. Tobón's friends were weaker than Hernández's friends, and Tobón died as a consequence. Friends and enemies mattered.

Ideas

By definition, imagined communities are based on ideas. For cotton textile workers, these ideas were deeply rooted in their real social relationships at work, at home, and in their physical communities. To further understand the workers' revolution, it will be useful to look at three sets of ideas that shaped their sense of community (race and nationality), their goals for the future (education and religion), and their sense of themselves (good times and bad).

Race and Nationality

In 1921, Chester Jones, "sometime professor of Political Science in the University of Wisconsin," remarked that "Mexico has recognized that her greatest problem is at bottom a race problem."[116] Throughout the colonial period, Spaniards saw themselves as white and dominant and Indians as dark and inferior. These beliefs reinforced colonial control, exploitation, and on occasion, atrocity. The long, slow, but profound acculturation of Indians and other subordinate peoples to Spanish ways did not change the fundamental perception that white was superior and dark inferior. In 1833, Esteban de Antuñano had his weaver affirm that "It is true, señor, the poor little Indians, one should look at them with great pity."[117] Eighty years later, José Fernández Rojas wrote that Zapatismo was "a problem in the South because of the indigenous race mixed with the zamba, because the population of Morelos, in its majority, is a cross of black and Indian, and the ideas and sentiments of this fearful human subspecies, blows like a hurricane in the rough, mysterious and dark spirit of the Indians of the states of Mexico, Puebla, Oaxaca and Guerrero."[118] In 1922, the victorious revolutionary government published an article by Francisco Escudero that stated that culture "in the Indian race is almost non-existent."[119] From Antuñano's "poor little Indians" to Fernández Rojas' "*temible subespecie humana*," to Escudero's "*casi nula*" Indian culture, there is not much distance.

In 1910 Nevin Winter wrote that,

Fully one-third of the population of Mexico are full-blooded Indians and another one-half are mestizo, those of mixed blood . . . Most of them are in about the same category as the southern negroes—a race without ambition. Content to be the servants of another race they neither court nor welcome change . . . A natural laziness, ignorance and a lack of interest will probably always keep down the peon's efficiency as a worker.[120]

Jones added that "it is the general testimony that the native lacks powers of sustained attention and industry. He is easily diverted from the task in hand. He shows, in short, in the work that he undertakes, an immaturity of character comparable to that of a child. These characteristics are emphasized in the hot regions."[121] He noted that mestizos,

"Whether through faulty education or other causes, the mixed bloods, up to the present, have not shown themselves an industrially able population . . . it is clear that Mexican labor problems will be in the future, as they have in the past, to a large extent, race problems."[122]

Thus Mexicans and foreigners alike held racialist ideas about Mexico's population. Mexican textile owners repeatedly stressed their French and Spanish origins, equivalent to stressing their whiteness. Fernández Rojas spoke for the country's elite when he wrote of the fearful dark underclass. While undoubtedly there were exceptions, it is hard to believe that most owners did not look down upon their workers for reasons of race, just as Escudero looked down on Indians. It is equally hard to believe that Mexican textile workers did not resent the racialism of the owners as well as of the foreign supervisors.

On the other hand, in the large mills most workers were male, skilled and literate, and often from urban families. Most of these men and women came from urban and often textile backgrounds. Whatever they felt about the owners, they most certainly did not consider themselves "Indian." The documents of the workers' revolution reveal very little about workers' racial feelings. Instead, one finds numerous references to nation and nationality. If owners sought "to conserve their national and ethnic identity,"[123] workers commonly noted the foreignness of any owner or supervisor they disliked. While the broader Mexican revolution had elements of an uprising of poor, dark people against their white and foreign masters, comments in the mills concentrated on nationality. When workers assaulted the owner of the Covadonga mill, the Mexico City newspapers labeled him "the Spanish Capitalist."[124] When the wealthy Spaniard reported that poverty drove the workers to attack his property, he put poor, dark, working Mexicans on one side, and a wealthy, white, foreign owner on the other. In 1920, Bernardo Barrolo, the El Léon administrator, told the local judge that workers directed physical attacks and death threats against "Spanish employees."[125] Some years later Vicente Lombardo Toledano wrote that in Mexico, "the directing minority has always been European race or mentality . . . Spaniards from Spain constituted this minority."[126]

After the 1917 Constitution, the Santa Rosa union referred to labor's legal victory as "the privileges that the law concedes to the Children of Mexico."[127] Nationality infused the workers' revolution as it did the broader Mexican revolution, but one wonders to what degree it represented intense if sometimes ambiguous racial feelings.

Religion and Education

Workers were unambiguously Mexican but ambiguously Catholic. As the owners noted, "the majority of workers profess the Catholic religion."[128] All the large factories and most of the small ones had their own chapel, which one owner declared "very necessary."[129] In that mill, Cocolapam, the chapel was dedicated to the Virgin of Guadalupe, a symbol of Mexico as much as of the Catholic Church. Despite their professed religion, many men rejected the Catholic Church hierarchy. During the Revolution, workers often occupied church buildings, using them for other ends. In Orizaba, the Cámara del Trabajo established its offices in the former "Catholic church of San Juan de Dios."[130]

Furthermore, their Catholicism coexisted with an unshakeable faith in what today we would call modernity. Driven by a belief in progress, workers fought vigorously for educational opportunities for themselves and their children. Their educational goals were part of their broader social struggle. During the labor uprisings of 1912, union leaders in Atlixco demanded that Metepec replace an old and inept schoolteacher.[131] When a federal labor inspector visited Puebla textile factories that year, he reported that "I could appreciate the anxiousness they [workers] have for instruction, both general and specific to the industry in which they work."[132] In December 1914, the head of the Cerritos union accused the owners of violating the education clause of the 1912 contract, claiming that the factory refused to "develop education."[133] A few months later, Cerritos workers wrote the Labor Department to ask for help with their school. Such demands were a constant of the workers' revolution.

Industrial workers not only wanted better education for their children, they also wanted it for themselves. In 1923, Veracruz textile workers asked for "night classes for workers, with the goal of achieving progress (*el adelanto*) among the working class."[134] In 1927 the carpenters' union of Tampico requested help to build a library "to help the intellectual development of our union members."[135] In the mill towns, the unions pressured industrialists to support workers' cultural centers.[136] The victory of their revolution led to the establishment of these centers in the mill towns, providing night classes and other educational and cultural events. Mill hands saw them as an expression of their "revolutionary principles" and a means toward their "general emancipation."[137]

Whenever religion and education came into conflict, it was usually education that won. Shortly after the revolution, textile workers at El

TABLE 3.9

Literacy Among Male and Female Textile Workers

Selected Mills, 1921

Mill	Men		Women	
	can read	cannot read	can read	cannot read
La Azteca	7	13	2	3
La Carolina	403	400	6	15
La Estrella	79	36	83	65
La Fama	263	80	99	22
La Perfeccionada	80	64	242	259
La Providencia	20	0	25	0
La Union	39	6	37	17
San Antonio Abad	233	153	67	61

SOURCE: Cuestionarios para el censo industrial, AGN, DT, Caja 288, Exp. 6. For La Fama, Cuestionario, Caja 298, Exp. 13. For La Providencia, SICT, Cuestionario sobre Trabajo, March 11, 1920, AGN, DT, Caja 207, Exp. 30.

Léon wrote a bitter letter to the municipal government demanding that it remove the local primary school teacher for teaching "our young children religious doctrine."[138] Thus Catholic textile workers were also modernists with a fervent belief in progress, ideals of education, and distrust of the Church.

Consistent with their values, most textile operatives were literate.[139] The chart above shows literacy in a sample of large and small mills. In the larger mills like La Carolina and San Antonio Abad, most men could read. Among women in the smaller mills, literacy varied but was usually less than in the larger establishments. Regardless, most desired the ability to read and write, if not for themselves, certainly for their children.

Good Times

Given natural individual differences, textile workers had clear ideas about good times and bad, which mostly involved social activities. Since their sociality formed the basis of their labor solidarity, it is worth describing in some detail.

In the male world of cotton textiles, the men shared their pleasures: alcohol, drugs, baseball, gambling, music, dance, theater, and sex. They worked long hours for little pay, but they enjoyed themselves as best they could. Outside the joys of family, most good times were spent with friends, drinking, carousing with prostitutes, and gambling.

Many men formed small bands to play popular music, while others contented themselves smoking marijuana, inside the mill and out.

Working-class life and alcohol came together. The mill towns and working-class neighborhoods in the cities were filled with cantinas and pulquerías. Wealthy foreigners complained that, "the habit of drinking to excess is the ruination of the working class."[140] Local authorities were more tolerant. They did not regulate drinking, although they did regulate cantinas and pulquerías. Municipal regulations usually prohibited women from entering, working in, or owning them. As Carlos Aldeco, the Jefe Político of Puebla argued, "It is completely proven that when women serve pulque, it gives rise to arguments, fights and scandalous behavior."[141] Pulque was also a staple of daily life.[142] When the Labor Department calculated the cost of living in 1914, it included pulque as a basic foodstuff for a family of "two women and a child."[143]

In addition to pulque, some men and women smoked marijuana. One textile worker remembered that, "In the barracks, those that had a woman, they would bring them food, and the head of the guards would stick a spoon in the pot and stir it to see what was inside; but the women were smart and brought a bottle of aguardiente in their breasts, and those that used marijuana, the women stuck them underneath the *paparrucha*; and while they talked, 'here is the package.'"[144] The use of marijuana was glorified in the revolutionary song, La Cucaracha, which celebrated the most famous cockroach in Mexican history, "the cockroach, the cockroach cannot walk any longer, because it doesn't have, because it is missing marijuana to smoke"[145] Textile owner Rivero Quijano attributed the nasty deeds of revolutionary military commander in Puebla, Coronel Silvino García, to marijuana, among the many who smoked dope during the revolution.[146]

Whether drinking or smoking, mill hands, like other Mexicans, loved music. They frequently formed bands or musical groups, some for the moment, others that lasted years. Francisco Reyes López, who began work in Cocolapam in 1908 at the age of 12, reported that he and his friends "organized a quartet . . . we played our small but select repertoire . . . and if they gave us a tip, it was the will of those who arranged the propaganda."[147] As unions acquired a more formal existence after the revolution, some of these informal groups became the official musical group of the union. In 1920, the Río Blanco union signed an agreement with "the members that form the Musical Band 'Free Workers' of Río Blanco."[148] Five years later, a workers' band asked the Atlixco Cámara de Trabajo to provide a monthly stipend "with which the working element can contribute to sustain the Musical Band in this town."[149] By the early 1920s, every mill town, every large union, and large mill had an official workers' musical group.

Alcohol and music were great but sex was best. Mexican women and men believed in family above self but this did not prohibit the men from enjoying sex outside the marriage, whether liaisons or lengthy affairs, either of which could lead to a second family. Gruening noted that "male conjugal fidelity among the Mexican upper classes is rare. Infidelity is flagrant and reputable. Mexicans speak without restraint of their children and their 'natural children.'"[150] Commenting on the poor, he stated that "Among the masses drunkenness and sexual debauchery form the aftermath of many religious festivals, and always have."[151] Gruening also claimed that his Mexican friends reported to not know "a single married couple where the husband was faithful to his wife."[152] Whatever his personal attitudes, Gruening was an astute observer.

Mexican workers had less money than the upper classes but it is not clear they had different sexual mores. Sex with prostitutes was an accepted part of male life in the mill towns. As with pulquerías, municipal governments regulated the activity. The 1902 Puebla code stated, that "Not being possible, without grave inconveniences, to avoid prostitution, the authority tolerates it, and therefore subjects persons dedicated to that activity, to the strict compliance of the articles of this code."[153] The code contained six categories of prostitutes, three for women who worked in "a public house" and three for women who worked alone. An amendment in 1915 required an inspector of sanitation to check the women's health.[154]

Even though prostitution was regulated, neighbors were not always happy with the activity. Prostitution often brought parties and wild behavior. Friends would often go out to enjoy a raucous night on the town, leading neighbors to complain, as did those in Orizaba in March 1920, about the "scandals" at the "house of prostitution" owned by María Victoria. They claimed that the men kept them awake at night, strolling down the street with their wild women, screaming obscenities and "shooting into the air."[155] With the men armed during the revolution, what could be more fun than hanging out with the guys to drink a local brew or smoke some dope, dance to a local band, enjoy sex with a working woman, and at the height of excitement, fire one's pistol? They were dangerous times, and they were good times.

Prostitution was ubiquitous, in the big cities and in the mill towns. Wherever there were factories, men, and money, there were prostitutes.

Although Mexico City was notorious as a center of nightclubs and cabarets, legal and illegal houses of prostitution flourished in Orizaba and Atlixco. The Mexico City government listed 9,742 legal women prostitutes in 1906.[156] Proportional to the population, there may have been even more in Atlixco and Orizaba. In 1920, the Orizaba Health Commission investigated clandestine houses of *"mala nota"* and *"muchos escándalos."* The Health Commission uncovered nineteen houses of prostitution in the Municipio, noting that there were others not yet registered.[157]

The owners of the houses were almost always women. Prostitution was undoubtedly degrading for many women, but for others it represented an activity that allowed some to become rich and important. As a business, however, it had its dangers. The women's worst enemy was the police. A later report on the activity noted that the police often behaved like gangs that exploited the women for their own benefit. "Our police are not distinguished . . . for their honesty."[158]

Unlike prostitution, gambling was often illegal. Labor inspectors frequently levied fines on workers who they discovered engaging in "small and large bets and other games prohibited by law."[159] Nonetheless, in every mill town and working-class neighborhood, gambling was as ubiquitous as pulque and prostitution.

Of course, not all textile workers hired prostitutes, got drunk and smoked dope, and gambled away their salaries. But most textile workers engaged their friends in the shop floor and in the community in activities that cemented the bonds of their layered communities. These bonds that emanated from daily life were the necessary social glue for the labor solidarity with which they built their revolution. Not all was pleasure, but all was necessary.

Bad Times

Not all was pleasure, because where there were good times there were also bad times. In addition to the normal difficulties of life, the violence of a prolonged revolution exacerbated life's precariousness. Murder became common, whether of rivals or enemies. A culture of male honor, alcohol, competition for women, political and union rivalries, and just plain fights often led to tragic outcomes.

"Very good friends" Agustín Medina and Luis Orozco worked in the San Antonio Abad mill in Mexico City.[160] They had an argument at work and after an exchange of angry words, a fight ensued. Medina hit Orozco, who then grabbed an iron bar and clubbed Medina in the head until his brains poured out. The police arrested Orozco for the murder of his former friend.

If work led to fights, so did alcohol. In early 1920, Manuel Ordáz, a foreman in one of the Orizaba mills, was having a drink in an open cantina when mill hand José Jayme walked by. We don't know the background, but just seeing the workman led Ordáz to proffer insults. When Jayme responded in kind, Ordáz smacked him with a mug, setting off an inconclusive fight between the two. The next day, Ordaz reported his version of the fight to the union's secretary general, who ordered the police to arrest Jayme on his lunch break. When the police released him, Jayme discovered that the union had fired him from his job in the mill "for not accepting a drink with the foreman."[161]

This fight reveals some of the changes wrought by the revolution in the relationships between friends and enemies at work. Although alcohol and fights were common enough before the revolution, the success of the workers' struggle created powerful unions that brought together foremen and workers. Furthermore, these unions could now hire and fire anybody in the mill, even intervening in fights and friendships, which is what cost Jayme his job. This was not a unique case. Francisco Reyes López, who we mentioned earlier playing music in a quartet, had a fight with another mill hand. The blows by Emilio Aguilar sent Reyes López to the hospital and Aguilar to jail. Aguilar sought revenge and accused Reyes López of being an enemy of the union, a *"contrario."* Normally the union would have expelled Reyes López immediately but his friendship with one of the leaders, "who used to be in a musical band with us," allowed him to present a defense. Reyes asked sincerely, "since when is the right of friendship prohibited?" Although the mills had always been networks of friends and enemies, after the revolution powerful union leaders intervened in these networks to achieve their own aims. The town's Labor Council expelled Reyes López from the union, so that he also lost his job.[162] Afterwards, the poor mill hand and amateur musician lamented that "they say that not all in life is glory, and that is the truth."[163] Not all in life is glory, nor in revolution.

Conclusion

Raymond Williams wrote that to understand the relationship between the individual and society, we seek "the essential relation, the true interaction, between patterns learned and created in the mind and patterns communicated and made active in relationships, conventions and institutions. Culture is our name for this process."[164]

The culture of cotton textile workers developed from eighty years of work in mechanized mills of increasing size, and eighty years of living in working class towns and neighborhoods. The given social relationships of work and daily life created in their minds interlocked layers of communities through which they defined themselves, their revolution, and their goals. These provided a nuanced concept of social class. To many, the idea of class may have encompassed violent opposition to class enemies, even the right to murder supervisors, yet retained room for a continuing hierarchy that they perceived as natural and to a certain degree, positive. The latter suggests the degree to which their communities shared ideas with the larger society, including the centrality of family and personal relationships, as well as a growing belief in progress. But what stands out among the workers is the degree to which the shop floor, where they spent most of their waking lives, shaped an intimate community of friends and enemies, solidarity and hatred, and ultimately, their peculiar view of social class. They worked long hours in the mills, and in the course of revolution, that is where they directed their attention. Overwhelmingly, the goal of the workers' revolution was to change the conditions and social relations of work, leaving most of the larger society to itself.

It was therefore the real experience of work that drove their conception of community. Mexican cotton textile workers were relatively uninformed by a European intellectual tradition, whether Marxism or anarchism. There were active anarchists, to be sure, as John Hart has demonstrated.[165] Nonetheless, a view of their lives at work strongly suggests that the practical ideology of the Mexican working class was determined by daily experience rather than by political parties, formal ideologies, or external leaders. Ultimately that is why their revolution, to which we turn in the next chapter, chose for its main organizational vehicle the trade union, which represents the workplace instead of the society-oriented political party, as in Russia. Mexico's workers' revolution was a trade union revolution because cotton textile workers learned community primarily on the shop floor and secondarily from their working-class communities. From there, they generalized to the rest of the world.

CHAPTER 4

The Beginning of the Workers' Revolution, 1910–1912

THE DEVELOPMENT OF COTTON TEXTILES through the nineteenth century created a factory proletariat that believed itself an oppressed class, enjoyed mechanisms of social solidarity, and strongly identified with family and community. It also created a core of workers, skilled weavers, who numerically dominated the larger factories and who understood their centrality to the industry. What it did not create was the idea of revolution, which came to textile workers from the outside. This chapter describes the opening of the workers' revolution in the context of Mexico's larger revolution. Because the workers revolt initially challenged work rules, the chapter begins with the evolution of work rules before the revolution. It then moves to the actual opening of the workers' revolution, the 1911 general strike, which led to an agreement to carry out a convention to solve the issues that caused labor strife. In the absence of repression, mill hands radicalized their activities between the close of the strike and the opening of the convention. Following a discussion of that meeting, the chapter concludes with its protocontract and labor's response, a response that became a challenge to authority at work, the subject of Chapter 5.

Cotton textile workers launched their revolution on December 21, 1911, only thirteen months after Madero's call to arms on November 20, 1910. The proximity of the dates and the parallel challenges to the social order were not accidental. Madero sought political change without a fundamental transformation of the social order. Textile workers initially sought traditional laborite goals, higher wages and lower hours of work, without planning to overthrow the labor regime.

Madero miscalculated the profound anger among much of the under-class in Mexico, along with the resentments of some in his own class. Those forces soon overwhelmed him, leading to his assassination in early 1913 and the subsequent civil war/social revolution whose consequences defined Mexico for most of the twentieth century. Similarly, only their unexpected victory in the 1911 general strike allowed textile workers to think, for the first time, that they could get more than their early and modest goals. The subsequent lack of effective repression caused by successive collapses of the central government—Díaz, Madero, Huerta, and finally Carranza—permitted them to carry out the unthinkable, to strike without getting fired, to attack supervisors, and finally, to demand control over the shop floor. As in the broader revolution, empirical process carried the revolution much further than its early protagonists had imagined.

Of course, it was the larger Mexican revolution that made the workers' revolution possible. One led to the other. It is inconceivable that Mexican industrial workers could have launched their revolution outside the context of a larger social revolution. Despite working in Mexico's largest and most advanced factory industry, cotton textile operatives by themselves were too small and too weak to lead a revolution in an agricultural and rural country whose wealthiest sectors lay in oil, mining, and export agriculture. Mexico was not Russia, and the Mexican proletariat was not part of a larger European proletariat that had long been versed in Marxism, socialism, and anarchism. Unlike their contemporaneous Russian counterparts, the early goals of most Mexican workers were too modest to lead a peasant revolution, though not so modest that they couldn't later take advantage of it. Furthermore, all previous strikes and workers' movements in Mexico either failed completely in the face of owner and government opposition or, if they succeeded in modest goals, failed to enact any fundamental transformation in the factory. It was the context of a larger social revolution that made a radical workers' revolution possible.

For workers, the most important consequence of the larger revolution was that it crippled repression from above. The collapse of the Díaz government in early 1911 quickly weakened the authority of local officials. It also undermined the capacity of the central government to militarily repress a growing workers' movement. It is suggestive that the interim government of Francisco León de la Barra, a former minister in the Díaz government, proposed the first federal labor office in Mexican history. Without the old dictator, Mexican elites realized that

they would have to do something radically new with workers: negotiate. Once Madero became president, he was too weak to repress
workers, or *campesinos* for that matter. The later collapse of the Huerta
government led to a period in which no central authority ruled Mexico. Many years would pass before government could put labor in
its place.[1]

The absence of effective repression, the importance of which cannot
be overstated, allowed three things to happen that explain the subsequent workers' revolution within the revolution. First, it allowed Mexican industrial workers to act on their anger. As seen in the previous
chapter, mill hands saw themselves as an underclass mistreated by
those above them. They hated arbitrary authority at work. The mechanisms of social solidarity among them were strong. But they were politically weak before the revolution and unable to transform their
feelings into a successful social movement. The outbreak of the larger
revolution and the consequent decline of effective repression allowed
them to win labor struggles, beginning with the general strike. Early
victories encouraged more strikes and labor actions, and victories
in these strikes and actions emboldened the more radical workers.
Worker self-confidence is what transformed early strikes over wages
and hours of work into work actions for control of the factory. Moderate walkouts became violent assaults on owners and supervisors. Incipient factory unions grew into national alliances and then into labor
confederations almost as powerful as the new state that emerged from
the revolution. Each failure of repression, each unpunished act of violence, each successful strike radicalized mill hands. When they reached
limits, they were of their own doing rather than those imposed by a
weak state. Mexican workers were less clear about private property
rights than Russian workers, but that is an issue we shall discuss later.
For now, what can be seen in the opening of the workers' revolution is
an absence of repression that emboldened workers. Once they entered
the terrain of social change, there was no external force to stop them.

Second, because workers had experienced trade unions before the
revolution, they turned to them when the opportunity to challenge owners arose. Because the unions did not lose the early strikes, workers
gained confidence in them, so that trade unions became the principal institution of the workers' revolution. Consequently, mill hands saw little
need to turn to clandestine political parties or workers' councils. From
beginning to end, the workers' revolution in Mexico was a trade union
revolution. Ultimately, however, if lack of clarity on private property

rights imposed limits to the workers' revolution, so did dependence on trade unions.

Third, the absence of effective repression forced owners to negotiate with workers, something they mostly did not consider during the Porfiriato. If owners came to believe in negotiation, so did the emerging trade unionists because trade unions require the state's legal apparatus to legitimate their existence. Thus both sides—workers and owners—came to believe that they could best achieve their goals by constructing a legal labor relations system. Owners thought new law might get the workers to behave, while trade unions needed law to recognize their right to be party to collective contracts, without which trade unions have little reason to exist. The workers' revolution quickly evolved into a struggle to create and control a new labor relations system. That is why law is so important to understand the workers' revolution, as we shall see in Chapter 6. For workers, the pursuit of labor law, like property rights and trade unions, became both a strength and a weakness of their revolution. Its strength was that it allowed them to win the most progressive labor legislation in the Western Hemisphere, which despite the precariousness of law in Mexico should not be understated. Its weakness was that law is a function of the state. The creation of new law could only strengthen a state that would in time first tame the labor movement, then control it, and finally move to crush its independence.[2]

While industrialization and industrial conflict in the late nineteenth and early twentieth century suggest that Mexico would have developed an institutional labor relations system with or without revolution, the workers' revolution determined that it would be the most proworker labor regime in the Americas. It began with a single strike over work rules, so it will be useful to describe the evolution of work rules before the revolution.

The Evolution of Work Rules Before the Revolution

In April 1848 Lino Romero sent Agustín de Antuñano, the son of Esteban, a set of rules for running his cotton mill. The *Prevenciones para el arreglo de la Economía* required *operarios* to work the full week or lose all their pay. From Monday through Saturday the workday began "from sunup" and continued "until 9:30 or 10:00 in the evening as is custom." At the end of each sixteen-hour day, mill hands had to clean the machines without pay or receive a fine. Romero recommended placing guards throughout the mill to prevent waste or theft.[3] The owners of the new industry demanded long hours for low pay in an environment of distrust and hostility.

The expansion of the industry in subsequent decades led to larger factories and more rigorous attempts to control workers. The owners relied heavily on the support of federal, state, and local governments. To prevent unionization, the federal Penal Code of 1871 provided for three months imprisonment and fines for those who "employ in any way physical or moral violence, with the object of raising or lowering wages and salaries of *operarios*."[4] Later, 1876 legislation required workers to obtain letters of good conduct from supervisors in order to apply for new jobs.[5]

As factories continued to increase in size, increasingly impersonal decision making increased labor conflict. Operarios, more frequently recruited from working-class rather than rural backgrounds, organized unions, demanded rights, and challenged supervisors and managers. Throughout the late nineteenth century, however, the mills regularly defeated whatever challenge to authority that workers threw at them. Workers lost most battles because they lacked strength and organization, their methods and goals were often timid, and the owners could usually count on local and national authorities to repress strikes if things got out of hand. In the 1870s, the operarios in El Hércules demanded cash payment instead of chips, but when they lost their struggle, 350 families had to move.[6] In 1889 the owner of La Carolina in Atlixco reduced wages. The workers struck only to find strikebreakers supported by the feared *Rurales*. One of the strikers ended up in prison while the others had to emigrate from the region. La Carolina, like other mills, restored order because it was able to fire rebellious workers, replace them with strikebreakers, and count on the armed forces to maintain order.[7] Workers who lost often had to leave town.

From Romero's work rules in 1848 to those posted in 1904 by Julio Gómez Abascal, for El Carmen, the owners' authoritarian control over the factory changed little. Administrators warned workers that they would be fined or fired for any imperfection in the cloth or any perceived violation of work rules and that "any appeal will be useless." Workers had no rights other than to earn a low wage for long hours of work. Factory administrators ran the mills as they wished and fired anybody who opposed them. It was an authoritarian regime based on Porfirian interpretations of private property rights.

Despite the owners' considerable advantages, textile workers carried out numerous strikes during the Porfiriato.[8] They fought over wages,

TABLE 4.1

Work Rules in El Carmen, 1904

Notice to *Operarios*

There will be no lending money, without exception nor for any reason.

The weaver who delivers an imperfect piece of cloth, will be fined because of the prejudice to the factory. Equally, all operators will be fined for any damages they cause.

The operator who enters the factory has the obligation to work the entire week at his assigned job, otherwise he will be charged fifty centavos for damage to the mill; he will also be fired from the factory unless there is some legitimate justification.

No maestro may fire an operator or give him permission to miss work without approval from the Administration.

These rules will be rigorously enforced, any complaint will be useless.

El Administrador
Julio Gómez Abascal

SOURCE: Julio Gómez Abascal, Aviso a los Operarios, January 1, 1904, AMA, Presidencia, gobernación, 1904, Caja 91.

the length of the workday, unpaid cleaning of machines, arbitrary fines and punishments, and authoritarian factory administrators.[9] The most famous strike took place at Río Blanco in 1906. In April, a small group of mill hands founded the Gran Círculo de Obreros Libres. When labor agitation spread to neighboring Puebla, local textile owners founded the CIM. The next month the CIM issued regulations for textile mills "which were intended to halt increasing labor agitation."[10]

The recently organized workers countered with a December strike in Puebla and Tlaxcala, to which the owners answered with a massive nationwide lockout on December 22. Porfirio Díaz tried to negotiate a settlement, albeit one that favored the owners rather than the workers. In Tlaxcala the governor sent the strike leaders to prison, effectively ending the movement there.[11] Throughout the country defeated textile workers sullenly returned to work, except in Orizaba. At Río Blanco, angry strikers confronted scabs entering the mill. Díaz then ended the Orizaba strike with a labor decree on January 4, 1907.[12] On January 7, troops intervened, "killing scores in the largest single massacre of non-indigenous people in the history of the regime."[13]

Although the literature has treated the 1906–07 conflict as a symptom of late Porfirian discontent, it was in fact a fight about work rules, increasingly inevitable in an industry of large factories. The owners' rules included the 14-hour day and six-day week, heavy fines for

TABLE 4.2

Porfirio Díaz's Work Rules, 1907

---

1. "Factories . . . will reopen on January 7, subject to regulations in existence at the time of closing . . ."
2. "Factory owners will continue the study . . . with the object of creating a uniform wage structure . . ."
3. "Every worker will carry a book in which shall be entered comments regarding his conduct, work habits, and aptitude."
4. Improvements:
    I. "fines . . . will be placed in a fund for the benefit of widows and orphans."
    II. "Discounts for medical fees, religious fiestas, or for any other reason will be eliminated. Every factory will hire a doctor for its workers."
    III. "Workers will only be responsible for materials and tools broken due to their negligence . . . this will be determined by the Administrator."
    IV. "Workers may receive visits from whomever they please, but must regard the rules governing good order, morality and hygiene."
    V. "When a worker is fired for cause, he will have a period of six days in which to vacate the company house, unless the cause of his dismissal was the discovery of arms, in which case he shall leave the same day."
5. "Workers with a grievance, should present the matter in writing to the factory administration, which will send a reply after fifteen days. The workers will be obliged to remain at work for this period, but can leave after receiving the answer, if unsatisfied."
6. "The factory will improve present schools."
7. "No children under seven . . . will be allowed to work, and older children only with the permission of their parents."
8. "Workers must accept the scrutiny of their journals and newspapers by the *jefes politicos.*"
9. "Workers will not be allowed to strike, least of all wild-cat strikes, since article 5 has established a grievance procedure."

SOURCE: Koth, 1993, 57–58.

flawed work, and a prohibition against allowing guests in the workers' homes on mill property.[14] On December 9, mill hands, who thought the CIM project "highly prejudicial for the interests of the workers *community*,"[15] countered with their own, marginally less onerous set of rules.[16] Their proposal included an end to the fines and such benefits as toilet paper in the factory bathrooms.

Díaz's decree of January 4 was neither contract nor law. It was perhaps closer to the responses of medieval Castilian kings than to modern labor codes, a specific solution to a specific grievance. In Atlixco it was ignored; Jefe Político Ignacio Machorro fulfilled his promise to labor leaders to not post it inside the factories.[17] In Orizaba the decree was implemented; widespread government violence had crushed workers' resistance to owners.[18] The decree became a government-sponsored set of work rules in some factories but not in others.

The nine-article decree, whatever its legal status, defined labor relations in the textile industry until the revolution. It provided support to the owners' iron-fisted control of the workplace, leaving intact their right to hire and fire at will and to run the factory as they wished. It required the good-conduct notebook for workers, which if not signed by supervisors could cost them their jobs. It absolutely prohibited strikes. It imposed government censorship of workers' journals. The owners recognized the support provided to them, because "thanks to the approved clauses, the *authority* of the factory owners remains intact."[19] There were some concessions to workers, particularly provisions for medical care and a loosening of the owners' control of the mill houses. But at the end of the Porfiriato, the formal institutional labor relations system was almost nonexistent. There was no body of labor legislation. Law recognized neither unions nor collective contracts, and labor organizations were barely tolerated, and strikes not at all. Government regulation consisted of the personal interest of local political bosses, state governors, or the president himself. In short, the labor regime was little formalized but what formality existed was that of the owners. Private property reigned supreme.

Their overwhelming power led owners and government officials to underestimate the depth of worker opposition to a one-sided labor regime. The official reports of the 1906 strike claimed that challenge to authority came not from workers but "from people who take advantage of the circumstances to profit from them." They argued that unions and strikes "violate rules and undermine discipline and respect for supervisors." They blamed "spontaneous strikes" on "professional agitators who carry on a nefarious labor," and "bad ex-workers or unscrupulous adventurers who desire to take advantage of the ignorance and lack of discernment of the mass of workers." They added that strikes taught workers to "disdain their superiors and to rebel against everything."[20]

Despite the opinion of owners, strikes emanated from the sense of community among workers. Between 1907 and 1910, textile workers continued to organize unions, demand better working conditions, and engage in periodic work stoppages. In Atlixco, the Metepec workers founded the Círculo Fraternal de Obreros on November 16, 1907, electing Panfilo Méndez vice president.[21] They carried out a strike in 1909 and two more in 1910.[22] In Mexico City, Puebla, and Orizaba there were significant strike movements in 1908 and 1909.[23] In 1908, the army intervened in strikes at the San Angel mills. In 1909, workers in San Antonio Abad carried out two strikes, though neither was successful and hundreds of mill hands lost their jobs.[24]

The 1909 Metepec strike was illustrative of the situation on the eve of revolution. On Monday afternoon, May 10, a group of weavers refused to return to work, demanding higher pay while protesting mistreatment by supervisors and "unjust antihumanitarian fines."[25] The next day the mill awoke to a massive walkout when the rest of the mill hands joined the movement.[26] While local authorities brought in the Rurales, the Mexico City offices of the Cía. Industrial de Atlixco sent a representative to negotiate. He partially accepted some of the wage demands, and the strike ended without anybody losing his or her job. Workers enforced strike discipline by harassing potential strikebreakers, contributing to their partial victory.[27]

Fortified by their success, Metepec workers struck again in January 1910 when the factory tried to lower their wages.[28] This time they lost, returning to work on February 3 after the factory fired the leaders and evicted striking families from the mill houses.[29] The defeat left the workers bitter yet unbroken. Two months later, Ignacio Machorro, Atlixco's Jefe Político, reported that labor militants "have been publishing subversive writings."[30] In October, just a month before the beginning of the revolution, Pedro Luna led another walkout.[31]

During this prerevolutionary period, the Flores Magón brothers and their Partido Liberal Mexicano (PLM) fought against the Díaz dictatorship. They espoused an anarchist ideology that offered workers a political language for their workplace grievances. The PLM's platform of July 1906 incorporated working-class demands: medical care, a national minimum wage, a six-day work week, industrial disability expenses, and standards for job safety.[32] The PLM played an active role in the events leading to the Cananea strike in the north in 1906.[33] Nonetheless, it is difficult to find their influence in central Mexico and in the textile industry. Rank and file actions suggest that many workers believed in and fought for working-class and antigovernment goals, with or without formalized ideology.

The 1911 General Strike

We do not know what industrial workers might or might not have achieved had Mexico experienced a stable political transition after Porfirio Díaz.[34] There was no smooth transition, but instead revolution.

Although Madero was a hesitant revolutionary, his followers were not, including many textile workers. The revolution broke out in Puebla, the center of Mexico's textile zone, and textile workers participated in the early Maderista movement and the subsequent fighting. Fifty-six Atlixco textile workers, mostly from Metepec, established the first Maderista political club in the state, with other clubs in La Constancia, La Independencia, Metepec, and La Tlaxcalteca.[35] When revolutionary leader Juan Cuamatzi raided Atlixco, it unleashed "a series of strikes and rebellions in several Puebla and Tlaxcala cotton mills."[36] By May 1911, the fighting in Puebla forced the owners to temporarily close many of the mills, "resulting [in] four thousand unemployed factory hands."[37] In Metepec, the mill hands joined the revolutionaries in looting and burning the factory, attacking foreign employees, and dragging one of managers through the streets with a horse.

When Díaz resigned the presidency in May 1911, many local authorities soon followed. This left local ruling groups to confront new and unprecedented challenges without traditional support.[38] Textile workers initiated strikes in Orizaba, Puebla, Atlixco, the state of Mexico, and Mexico City.[39] The causes varied from "bad treatment by a supervisor" in Río Blanco to "terrible working conditions" in La Colmena and El Barron. The strikes demonstrated that mill hands had not been decisively defeated in the aftermath of 1906 and that they could mobilize if the state was unable or unwilling to repress them. In some factories, mill hands engaged in direct violence. When the victorious Madero planned a visit to Puebla, workers in La Covadonga attacked and robbed the factory, killing several foreign employees.[40] Workers launched four separate strikes in Metepec and El León in the first three months of 1911.[41] In the first strike at El León, weavers demanded the firing of their supervisor. This was followed by a similar demand and a strike by weavers in Metepec. In the second Metepec strike, male weavers walked out when the mill hired a woman to review their work.

Labor conflict continued throughout the year. At Río Blanco in early September 1911, a mill hand wounded a supervisor when he hit him in the head with a machine piece. The administration fired him, enraging his comrades who shut down the mill then took to the streets in an angry mob. Instead of fleeing at the sight of the formerly feared Rurales, they fought back with stones, wounding a sergeant. The mob then attacked and destroyed the house of the supervisor Julio Vivar. Only quick action by the troops saved Vivar from being murdered.[42] A frightened municipal government quickly made a number of arrests, banned all meetings of more than ten persons, and denounced the workers "subversive speeches."[43] The Río Blanco union, Solidaridad Obrera, threatened to strike and forced municipal authorities to arrest the wounded Vivar.[44] Though a few more arrests were made, both local authorities and company officials remained fearful of "new disturbances."[45]

Under Díaz, it would have been unthinkable for the Rurales to have been attacked by workers without extreme repression. Now, with revolution, angry workers successfully attacked Rurales, the factory administration, and the town government, suffering only a few arrests. If events in Chihuahua and Morelos demonstrated the willingness and capacity of campesinos to pursue revolution, the strikes at Río Blanco and Atlixco suggested that textile workers also had the energy and desire for radical social change.

Instead of taking office in November 1911, to the applause of a grateful nation, Madero confronted armed rebellion in Morelos, resistance to central authority in Chihuahua, and just weeks later, "the first labor action of great importance in the Republic."[46] On December 21, 1911, textile workers in the fifteen most important mills in Puebla and Atlixco went on strike.[47] The movement quickly spread to Tlaxcala and Mexico City, shutting down the entire industry with the important exception of Orizaba. The leadership, the *Mesa Directiva de la Huelga*, came from workers and union activists in the old La Constancia mill in Puebla. They demanded a reduction in the workday from 12 to 10 hours and a "reasonable" wage increase."[48] The mill hands claimed that their demands were "of strict justice . . . within the limits of the rational."[49] Strike leaders reported to Madero that:

The workers of the textile factories of the State of Puebla, we didn't declare the strike simply out of a desire to earn more. The entire strike began in a single factory because the administrator or the manager ordered that workers without families could not eat in the houses of others who voluntarily agreed that their families provide them food, something absolutely indispensable, because without this help, workers without families would not have a place to eat, because in general the factories are not located in towns but some distance away, which impedes single workers from having a place to eat. This was the beginning of the strike, because the workers did not want to be bothered in matters in which, in reality, the factory was not hurt.[50]

Three points are important to note about the general strike. First, its public goals were the traditional laborite demands of wages and hours of work. Second, it was preceded by great violence in the mills, with workers physically attacking supervisors, particularly foreign supervisors, which suggested the willingness of workers to move beyond traditional goals. Third, workers never lost sight of community issues such as gender and food. Male workers did not prepare food, which they considered women's work. Furthermore, workers could not accept factory interference in a matter in which "the factory was not hurt." In different mills, workers added other grievances, including a change in cloth size in La Hormiga, unjust treatment and stripdowns in La Carolina, the right to receive friends and relatives in factory-owned houses in Metepec, and dignity, or as the workers said, "better treatment by the owners."[51]

In Puebla, Porfirian governor Mucio Martínez resigned office on March 2, 1911, replaced by interim governor J. Rafael Isunza, who was then replaced by Nicolás Meléndez on December 25. Taking office in the middle of the most important strike in the state's history, and with only uncertain support from Mexico City, Meléndez assured the local Congress that he would "improve social conditions for workers."[52] When he met with workers three days after taking office, he promised to arrange a meeting of the owners and the strike leaders.[53] Two days later, the industrialists, led by "the Spanish Consul, don Manuel Rivero Collada," met with the new governor. That afternoon, three hundred textile workers invaded the governor's palace, wishing to speak directly with Meléndez.[54] The new revolution became a civics class for workers, teaching them the new rules of politics: direct action.[55]

The textile strike reverberated throughout a country in which the old mechanisms of political control were collapsing. Rural laborers carried out strikes against haciendas in various parts of the country.[56] In early January, miners, textile workers, and farm workers in Chihuahua, Jalisco, and Tlaxcala joined the textile strike with their own demands for higher wages and reduced hours of work.[57] Operarios in the Bajío also walked out.[58] Although some upper-class Mexicans had originally supported Madero, rapidly growing social strife convinced many of them that the revolution was an assault of the poor against the rich. Angel Díaz Rubín, the owner of La Covadonga, claimed that his mill hands had even looted his ranch, Moralillo, going so far as to steal the doors.[59]

Confronted with escalating violence, the Puebla owners met at the Centro Industrial Mexicano on January 5. They agreed to reduce the workday by one hour but refused to grant a wage hike, arguing that they already paid the highest wages in the country. They also refused "all the other petitions of the operators."[60] Concerned about a wage squeeze between workers who wanted more and regional competitors who paid less, they placed their hopes "in a general meeting to take place in Mexico City, soon, to which all the industrialists of the country will attend. It will attempt to unify wages in the Republic."[61] Mill hands rejected the peace offering.[62] In mid January, the leaders of the Orizaba textile workers, who had previously stayed aloof from the conflict, delivered a list of their own demands, including an end to fines and blacklists, recognition of unions, and limits to at-will firings.[63] In fact, they demanded a reverse fine in which the mills would pay workers if a faulty machine stopped working.

President Madero, just weeks in office, now confronted the most serious challenge to the owner-dominated labor regime in Mexican history. Although he neither wanted nor planned a widespread social revolution, he lacked the power that Díaz had employed in 1907 to resolve a strike through repression. Unsure of his military backing, Madero turned to the mediation proposed by the Puebla industrialists. On January 9, Gustavo Madero, the president's brother and minister of government, invited the owners of textile factories to a meeting to be held in Mexico City on January 20, the purpose of which would be to discuss a standard wage scale throughout the industry and the reduction of the workday to ten hours.[64]

On January 20, 1912, Rafael Hernández, Ministro de Fomento, convened the meeting. The country's most important textile industrialists attended, including Adrian Reynaud (El León), Ernesto Sánchez Gavito (La Beneficencia), Tomas Reyes Retana, Enrique Artasánchez (El Volcan), Manuel Morales Condes (La Trinidad), Santos de Letona, Adolfo Prieto (La Victoria), Angel Díaz Rubín, and others. With Gustavo Madero in attendance, Hernández's opening address made it clear that the widespread strike movement had forced the government to enact legal reforms:

Because of the country's recent political change, there have developed a number of social movements. Strikes have followed upon strikes, and the government, worried about this state of affairs and wishing to find a solution, agreed to convoke this meeting with the object of discussing the bases to regulate labor in the Republic, which the government proposes to elevate to the category of law, presenting initiatives to the Federal Congress."[65]

Hernández's explanation of the beginning of the workers' revolution is coherent. Madero's overthrow of Díaz opened the door to social change, leading to massive unrest in the textile industry, which then ceased to function with the success of the general strike. Both government and owners felt incapable of repressing workers, so instead they turned to concessions, among them a willingness to construct a legal labor relations system, something completely new to Mexico. Following this logic and to "calm the worker situation and solve the strike,"[66] the owners signed a temporary agreement to reduce the workday to ten hours and increase wages 10 percent.[67] They also established commissions to study labor problems and wage uniformity.[68] With this apparent victory, labor leaders approved an end to the strike. Workers and owners agreed that the January 20 settlement was provisional until a later meeting of the industry drafted a final settlement.

The workers had won the first round.

From the Close of the Strike to the Opening of the Convention

The violence in the mill towns, the general strike, the January 20 agreement, and the plans for a first-ever convention of the industry forever changed labor relations in Mexico. Most important, the workers had won. They organized unions, carried out an industrywide strike, forced government to meet with the owners to resolve labor problems, and won more pay for less work. Trade unions and a militant strategy brought significant gains. Owners and government saw the obvious, which was that the old Porfirian repression was no longer sufficient to enforce an owner-dominated labor regime.

In the days after the strike settlement, there were warnings about the new labor climate. Angry owners, like those in San Antonio Abad, began to fire strike leaders "with the object of spreading terror among the others," with the result that sometimes "there is nobody who wants to speak in the name of the workers."[69] In other mills, such as San Lorenzo, mill hands expanded their union activities, such as collecting union dues inside the mills against factory orders. When administrators threatened them, they threatened back with strikes not only in their mill but at neighboring factories "because between both groups exist a solidarity pact."[70] In Atlixco, mill hands insisted on challenging the authority of factory work rules by entering the mills to work with their hats and sarapes on, collecting funds for striking workers in the region, and engaging in walkouts to protest firings, forcing the owners to threaten a regional lockout.[71] Local authorities arrested ringleaders.[72]

Everybody anticipated some sort of brave new world of labor relations. Government, owners, and workers each began to organize for the coming negotiations. The de la Barra government had taken the first steps toward institutionalization on July 24, 1911, when it proposed a Labor Office within the Secretaría de Gobernación, sending the *Proyecto de Ley* to Congress on September 22. Arguing that Mexico confronted no problem "of such urgent importance as that of labor" and citing similar offices in England (1887), France (1891), and other European countries, de la Barra proposed an office that would not only gather data but also "obtain an equitable arrangement in cases of conflict between owners and workers, and serve as arbitrator in such differences."[73] His projected budget included twelve positions, including a director with a daily wage of 18 pesos, two inspectors with daily wages of 8.50 each, and two *meritorios*, to earn 1 peso daily.[74] The Congress did not approve the Labor Office until December 18, after Madero had assumed the presidency and three days before the general strike broke out.[75] When the strike ended, Madero appointed Antonio Ramos Pedrueza first director of the Labor Office.[76]

The temporary settlement on January 20 spurred the institutionalization of labor relations. Ramos Pedrueza took credit for the idea of a tripartite convention that would bring together capital, labor, and government because he saw "the necessity of procuring the greatest number of supporters for the work rules, and to get the largest number of factory owners to adopt and use the wage scale."[77] It is not clear that participants foresaw the eventual outcome, but a national agreement to replace individual factory rules would inevitably function as a sort of industrywide collective contract. Because law lagged behind practice, the participants in the new labor negotiations did not talk about contracts but instead used the old terms, *reglamento de trabajo* and *tarifa de salarios*. Nonetheless, the January 20 agreements, by recognizing national standards, moved toward the idea of collective contracts.

The Labor Department noted that the strikers accepted the provisional accord "in the understanding that things would proceed from studying reforms to the wages scales and work rules and if possible towards the unification of both in the Republic."[78] For most workers, a new institutionalization of labor relations seemed beneficent. Lacking collective contracts, labor laws, and government offices, the old labor regime rested on the institutions of private property rights and the labor market, which favored owners. Mill hands needed something else.

Owners and workers agreed to establish separate committees to study wage and work rule unification. As a draft for discussion, the Labor Department translated the English Tarifa Minima Uniforme Inglesa, hoping that both sides would quickly agree to follow the English example.[79]

The industrialists, however, were not happy that workers had any voice at all. They agreed that the future textile convention would meet without direct worker participation, only allowing for a parallel meeting in which workers could react to the owners' proposals. This set in motion attempts by owners and labor leaders to organize their respective forces as both sides began to draft preliminary documents.

The Puebla and Tlaxcala owners already counted on the CIM. The CIM, however, had failed to defeat the workers in the 1911–12 general strike. Many industrialists wanted a newer, more aggressive organization and turned to leaders of the old political class to organize one. Chief among them was Tomás Reyes Retana. Reyes Retana had served as a federal congressman in the 1890s and later as a Porfirian senator representing Mexico City. A member of various legislatures under Díaz, de la Barra, and Madero, he was keenly aware of the changes in the political and labor climate. The representative of CIDOSA and several large Mexico City factories at the January 20 meeting, this aggressive and intelligent Porfirian lawyer and politician wanted an organization for the mills that he represented, which were larger and more modern than the CIM/Puebla factories.[80] On May 8, 1912, Antonio Reynaud (CIDOSA), Félix Vinatier (CIVSA), León Barbaroux (La Hormiga), and other powerful industrialists and lawyers from textile centers outside Puebla/Tlaxcala and the CIM, founded the Confederación Fabril Nacional Mexicana (CFNM).[81] The CFNM represented the French and Mexican industrialists who owned the large and modern mills in Orizaba, Atlixco, and Mexico City, as opposed to their competitors from the Puebla CIM, with smaller, perhaps less efficient operations.

Both ownership groups participated in the Committee of Industrialists, chaired by Reyes Retana, which proposed work rules on May 27.[82] This proposal varied little from Díaz's 1907 decree except for the reduction of the workday to ten hours, a measure agreed on in order to end the strike. Owners demanded control over the factories, including the determination of hours of work, at-will firings, and control over the mill houses. The general strike had not convinced them that a fundamental change had taken place.

While the owners not surprisingly agreed on work rules, they did not agree on wages. The CFNM represented larger, more modern mills that could compete favorably on costs and so opposed standardized wages. The CIM represented the smaller and less productive mills who wished to control cost competition and so favored a nationwide wage

TABLE 4.3

Industrialists' Proposed Work Rules, May 1912

"Proyecto de Reglamento Para las Fábricas"

Article 1—Ten-hour workday.

Article 2—Factory directors would determine when workers began and ended work and also lunch breaks, "according to the established customs."

Article 3—Workers had to be in their assigned place "without the right to go to another department nor interrupt the attention of another worker or workers for any motive or pretext."

Article 4—Prohibited drunk workers from entering the mill. Made workers responsible for the damages they caused.

Article 5—Workers had the obligation to work the entire week.

Article 6—During work hours, workers were to attend only to work matters.

Article 7—In case of a complaint, the worker had to present it personally or in writing to the department head or factory administrator.

Article 8—Factory-owned houses were exclusively for factory workers, who could not permit anybody to stay there without the consent of the factory owner.

Article 9—Prohibited maestros from demanding or accepting money in exchange for giving workers a position.

Article 10—Prohibited workers from smoking inside the factories. Prohibited workers from introducing matches, newspapers, alcoholic beverages, or weapons inside the mills. Allowed factories to fire workers for disobedience, insubordination, ineptitude or boisterous demonstrations. Abolished fines.

Article 11—Since the *reglamento* formed part of the work contract, laborers showed their acceptance by entering work.

SOURCE: "Proyecto de Reglamento para las Fábricas . . . ," AGN, DT, Caja 15, Exp. 18.

scale.[83] This disagreement portended conflict, even without taking into account the opinion of rebellious workers.

Madero's government, on the other hand, needed to appease workers. Having come to power through a revolution that toppled Porfirio Díaz and not yet having consolidated any sort of hegemony, the new government was more sensitive to the workers' rebellion. At the May 27 meeting of the Committee of Industrialists, Gustavo Madero suggested that the federal government might reduce its tax on the industry by 21.2 percent if the mills increased their wages that amount.[84] In effect, he proposed that government subsidize labor peace.[85] Reyes Retana publicly attacked the Labor Office for suggesting any wage hike.[86]

As convention preparations proceeded, the owners were divided amongst themselves and also with government. Labor also was not united, splitting naturally into the moderates, who wanted to work with government to consolidate wage and hour gains, and the radicals who wanted more. The more cooperative leaders of the recent strike formalized the Comité Central de Obreros de la República (CCOR) on February 12, electing an executive board of Benjamín H. Meza, president, Rafael Silva, secretary, and Vicente Estrada and Alfonso Reséndiz. Silva was a Labor Office inspector, suggesting government sponsored leadership. The CCOR prepared labor's proposal for the upcoming convention.[87] On March 26 they submitted a plan for a permanent settlement that included a uniform wage scale.[88] The national leadership expressed a belief in "the government's good will to solve things favorably."[89] They did not want the 60,000 or 70,000 workers they claimed to represent undermining negotiations through wildcat strikes.[90]

Rank-and-file workers were generally more radical, with greater confidence in themselves than in the good will of government. Their most immediate concern was to protect their nascent trade unions, which were local and not national and without contracts or legal safeguards. The natural leaders who emerged from the victorious strike were not afraid of conflict. In March 1912, the CCOR expelled José Otáñez, a leader of the December strike, as a "dangerous agitator."[91] Undeterred, Otáñez continued to "stir up the factories."[92] Owner Rivero Collado claimed that Otáñez told mill hands that owners "tried to exploit workers."[93] Ramos Pedrueza suggested to one of the industrialsts "that it will be very convenient that you or somebody influential with the Puebla governor gets Otáñez thrown in prison for some time."[94]

Even this plea revealed the sudden change in Mexican labor affairs. At one time, industrialists counted on effective government repression, but now government had to plea with them to help repress workers. Rank-and-file radicals took advantage of this weakness to carry out local actions. In March, the El Carmen operarios carried out a wildcat strike in support of their demand to choose who handed out the cloth, instead of the administrator.[95] In mid-June, the Santiago workers (Tulancingo) threatened to strike when the administration fired the union president and a dozen labor activists.[96]

The situation became even more dangerous when Orizaba workers began to show signs of militancy. When a department at Río Blanco closed its doors five minutes early, locking out the workers, the operarios forced a confrontation with the factory administrator. The administrator backed down rather than have a strike that would shut down the entire plant. Even before talking to management, however, the Río Blanco workers organized a meeting with representatives of the other

Orizaba mills. They promised another general strike if the industrialists who were about to meet in Mexico City did not accept their petitions.[97]

When the Industrialists Convention opened on July 2, Río Blanco, Cocolapam, and Santa Rosa workers finally walked out. Cocolapam had expelled a union leader from the mill, leading to a strike there as well as in Río Blanco, San Lorenzo, Miraflores, and Santa Rosa. At Río Blanco, the strikers tried to convince nonstrikers to join them. As the workers shouted "Long live liberty!" and "Death to the tormentors of the people!" the *voluntarios* hired to protect the mill surrounded the workers, somebody opened fire, and "scandalous workers" attacked an official, causing him to fall from his horse. While falling, he opened fire, a signal for other voluntarios as well as federal troops who were present to also begin shooting. Within twenty minutes, there were thirty dead.[98] The company blamed the workers, arguing that they initiated the violence.[99] Later reports suggested that many of the voluntarios were drunk and had chased after fleeing workers in sadistic glee.[100] In Mexico City, the convention needed no greater reminder of Mexico's new labor problem than thirty dead mill hands on the streets of Orizaba.

Thus on the eve of the Convention, the labor situation was radically different than just six months earlier. Workers, owners, and government all counted on new organizations to defend their interests. Even though there were divisions within each group, workers and owners were more militant than before. Workers felt a new confidence and were quick to pull the trigger on strikes and labor actions. Mill hands were fighting and dying on the streets in front of the factories. The owners sensed the danger and demanded the federal government save them. Madero walked a dangerous tightrope.

When Orizaba went out on strike, the workers won the second round.

The Convention

On July 2, 1912, Rafael Hernández, Madero's Minister of Development, inaugurated the much-awaited *Convención de Industriales*. Ramos Pedrueza wrote that it was a "memorable date in the annals of the Mexican industrial movement. For the first time the representatives of Capital and Labor have discussed all these points under the protection of a Ministry of State."[101] Among the approximately seventy delegates, every important textile owner was there or sent a representative. Some delegates, like Reyes Retana, represented more than one factory.[102]

Called to solve a conflict between owners and workers, the formal convention excluded labor. Nonetheless, by requiring the presence of workers at the convention, albeit in the hallways rather than in the meeting room, and by compelling labor's Central Committee to sign the final accord, the meeting was a de facto tripartite conclave among management, labor, and government. Since its basic goals were to write industrywide work rules and wage scales, it was essentially a tripartite convention to draft a nationwide collective contract, though Mexican labor law still lacked the terminology for such.[103] Thus the issues were new, difficult, and compelling. Owners, accustomed to command, believed a collective agreement would limit their rights and the rights of property. The costs of peace were high, perhaps too high. Moderate labor leaders needed an agreement that would satisfy an increasingly radicalized rank and file as well as opposing owners. They feared that *operarios* would not respect their moderate goals.

Making things worse, the two ownership groups fought incessantly. Convention rules called for each factory to have one vote and for the "definitive project" to be "subscribed by the members of the Convention and by the Central Committee of Workers."[104] Manuel Sánchez Gavito, spokesperson for the CIM, and Reyes Retana, leader of the CFNM, spent the first days of the meeting in a bitter fight over these rules, Reyes Retana seeking a proportional voting system based on the size and importance of each mill.[105] After some hardball negotiations, including a CFNM threat to boycott the convention, Hernández brokered a compromise.[106]

Meanwhile, on July 3, only two days after returning to work, the Puebla mill hands went forward with labor leaders' worst nightmare, a new general strike.[107] Most of them returned to work between July 17 and July 20, but some stayed out until August 10. The owners were "quite indignant because of the damages," fearful that government could not control the labor situation.[108] They continued to blame labor conflict on "agitators" and insisted that government officials arrest and imprison the troublemakers.[109] Rafael Hernández expressed the frustration of government and owners when he demanded that the workers' committee provide an explanation for the events in Puebla, where "all the factories are on strike."[110] Juan Olivares, an old weaver from Atlixco, a moderate and a leader on the workers' committee, had to acknowledge the division between the moderates on the workers' committee and a more radical rank and file.[111] He blamed the strikes on two former members of the committee, Méndez and Ramírez, who had returned to Puebla to promote strikes, adding that the owners' proposals "might have disgusted the Puebla workers."[112] In fact, radicals had returned to Puebla to tell workers that the leaders of the CCOR—Benjamín Meza, Juan Olivares, and Francisco Mendoza—had sold out to the industrialists.[113]

Even at this early stage, control over the factories, or as the owners saw it, the rights of private property to command labor, had become a central issue. On every dispute that might limit the power of factory administrators, the owners united in their recalcitrance. This angered even the moderates among labor, who complained that the owners reneged on their promises to cede something to workers, particularly the right to collect union dues inside the factory.[114] The rank and file demanded defense of their unions. In response, Sánchez Gavito claimed that the workers' demands were not "socialist but ultra anarchist."[115] Owner Martínez Carrillo added that "sooner or later the workers would see the factory as a place to arrange personal or family matters."[116] Another owner cut to the core: "workers could not and had no right to intervene in factory rules."[117] Meanwhile, a cynical Reyes Retana who had opposed the workers' desires, said that he "could not help but note that in this Assembly the industrialists have ceded a large part of their sovereignty and powers that are exclusive to them, generously and voluntarily," with the goal of helping the *"benemérita clase obrera."* He attributed to the owners *"un arranque de filantropía,"* noting that they had willingly accepted "the new socialist ideas."[118]

In fact, the owners listened little and ceded nothing not already lost in the January 20 agreement. On July 17, 1912, Ramos Pedrueza signed the new work rules for Mexico's textile industry. With two exceptions, they were not much different from the original May proposals or even from the 1907 Díaz decree. The new rules prohibited dues collections, imposed fines, allowed factory administrators to run the factories as they wished, provided for at-will firings, and while permitting workers to complain, allowed administrators to serve as judge, jury, and witness. Furthermore, they extended that control to the mill houses. The two exceptions were the reduction of the workday to ten hours and the establishment of a national set of rules, both a direct consequence of the general strike. There were some minor concessions: 50 percent overtime pay, formal holidays, raising the minimum age for minors to fourteen, and some educational and medical obligations for employers.

Even though the Reglamento provided for the control of owners, it was still a fundamental break with the past because it offered a new, written, formal, institutional labor relations system for the textile industry as a whole. Unlike Díaz's 1907 decree, this was not a unilateral

TABLE 4.4

The Reglamento de Trabajo, July 1912

Article 1—Set the work day at ten hours, nine for the night shift. Provided for 50 percent overtime pay.

Article 2—Factories set shift times, with work to begin between 6 and 7:30 A.M.

Article 3—"During work hours, each worker will stay in his place in his department, without distracting his attention on matters foreign to his obligations and will not go to another department unless his work requires it, nor will he interrupt the attention of other workers for any reason." Prohibited union dues collection inside the factory.

Articles 4 and 5—Made workers responsible for taking care of the machinery and for damages to them, providing for fines but also arbitration.

Article 6—Workers had to work the entire week.

Article 7—Mill hands had to present complaints in writing to the administrator, outside of work hours, with a resolution in ten days.

Article 8—Only the worker and his family could occupy the factory houses.

Article 9—"It is strictly prohibited for employees and *maestros* to mistreat the workers by word or by deed; to demand or accept money from them in exchange for work, or for any other reason. They are also prohibited from lending money at interest, or any similar abuse, under penalty or expulsion or both."

Article 10—Prohibited drunkenness and smoking, and also bringing matches, newspapers, alcohol, or weapons into the factory. "Workers will leave their hats and coats in the place designated for them, and can only enter the departments with small caps."

Article 11—"Acts of disobedience, insubordination, and lack of respect towards the administrator, employees, and other superiors, noisy demonstrations, and ineptitude, will cause the rescinding [of] one's work contract and the separation of the worker, without the necessity of a resolution by any authority."

Article 12—Abolished fines by replacing them with an indemnity to the factory in cases of "carelessness, ineptitude or wickedness" on the part of workers.

Article 13—Established Sunday and holiday rest, with holidays set on January 1 and 6, February 5, March 21, Thursday, Friday, and Saturday of Holy Week, May 5, June 29, September 16, November 1 and 2, December 12 and 25, and *"el día de la fiesta local del patrono de la fábrica."*

Article 14—Reglamento constituted a contract between worker and owner.

Article 15—Eight days notice to rescind the contract.

Article 16—Prohibited children under fourteen from working in the factory.

Article 17—Prohibited commercial monopolies and established payment and loans to workers in cash.

Article 18—Established the responsibility of owners to provide some education and medical care to workers for work accidents.

Article 19—Peons or specialized laborers to carry heavy materials.

Article 20—This Reglamento replaced all previous work rules; should be placed in a visible spot in the factories.

SOURCE: Contrato de Trabajo, AGN, DT, Caja 978, Exp. 3.

suggestion that some might follow and others ignore. This was an agreement signed by owners, labor, and government. The federal government would enforce it with punitive taxes on noncomplying mills. It was the first formal institutional constraint on the command aspect of the property rights of owners since they first built their mills. "This labor contract, the first in Mexico which had even a semblance of collective bargaining, seemed at the moment very radical in its provisions."[119]

Who won and who lost? On the one hand, it was overwhelmingly procapital. Workers had only limited rights. They could not collect union dues inside the mills. They still had to pay fines. Factory administrators still controlled the shop floor, even reading materials. On the other hand, the convention and contract represented real gains for workers. Through self-organization and struggle, they won more pay for less hours of work, uniform work conditions, mildly increased benefits, and written prohibitions against mistreatment. Since the state enforces contracts, the new Reglamento brought the state into labor relations. Furthermore, it subjected labor affairs to contractual negotiation and legitimated tripartite meetings as the mechanism to negotiate such contracts. This was indeed revolution, albeit limited in comparison to events outside the meeting rooms of Mexico City. Self-serving as always, Sánchez Gavito claimed that the owners "ceded to the pretensions of workers . . . and in this way, the industrialists reacted altruistically and with great nobility to the formidable national strike movement, which broke out lacking any justification."[120]

The industrialists then suspended their work for a week in order to study their next task, uniform wages. They then fixed wages for the smaller and less conflictive wool industry with an agreement on July 24.[121] The next day, cotton industrialists began their debate on the Tarifa Minima, using as an initial draft the "1911 Blackburn List of Prices for Weaving, as well as the Oldham spinning rates, adapted to our circumstances."[122]

By and large, owners and government believed that the market should set wages, not a central authority. Reyes Retana stated that the minimum wage was just that, a minimum that factories could exceed. He said that government should neither set wages nor restrict the rights of factories to compete for workers through wage differentials. The laws of supply and demand were, he said, "of natural character."[123] Ramos Pedrueza, the voice of government, added that "labor is a commodity subject to the laws of supply and demand, and as such, the workers will have lost their cause if they go against sociological principles that are like the physical laws that govern us, such as universal gravitation."[124]

Workers argued that wages were not set like the laws of gravity, and insisted on "making equal the wages of men and women."[125] The owners disagreed because, "the effort of a man is superior to that of a woman and produces more, so that a male wage ought to be superior to a female wage."[126]

After some discussion, the industrialists voted to accept the English wage scale, arguing that "the minimum wage in weaving is based on the English wage which is accepted by many thousands of workers and has been used for a number of years . . . and besides, our proposed minimum wage scale is 20% higher than the English . . . there is no reason that Mexican workers should demand more than what English workers have gained."[127] It was a strong argument that labor accepted.

The Tarifa de Salarios was a ten-page, detailed, industrywide wage scale. Because the industry mostly employed piece rates, the Tarifa specified job category, machine, and cloth type. The technical discussions during the Convention made it clear that it would require great expertise to manage the scale. For workers, this could only mean an increased role for unions. Although the wage scale included the union demand to not discriminate against women, wages remained a function of job category, which as we have seen was determined by gender.[128]

In theory, the two agreements, work rules and wage scale, were voluntary agreements among industrialists, enforced by a punitive government tax. In practice, it was as close as the country could get to an industrywide collective contract given the absence of that in law. In January 1913, Ramos Pedrueza stated that "using the wage scale is obligatory by law."[129] Weeks later, Adalberto Esteva, the new director of the Labor Department under Huerta, stated that "The Work Rules for the Weaving and Spinning Factories of the Republic, of July 17, 1912, is a bilateral agreement that binds both parties . . . the industrialist and the worker."[130] This was the beginning of an institutionalized labor relations system.

The closing session on August 1 was mostly self-congratulatory. The owners counted on a new alliance with the state and an incipient labor bureaucracy to control the rebellious working class.[131] They believed that the convention had drawn up measures "that for now were sufficient to solve the labor conflict."[132] Ramos Pedrueza noted that workers had spoken with the owners *al tu por tu,* pointing out labor's unprecedented gains: the ten-hour day, overtime pay, and arbitration for disputes. More important, "workers have been given a voice and vote with which to arrange their difficulties with the owners."[133] Even then, however, there were symptoms that all was not well when the labor chief informed the delegates that CCOR had asked for time to go to the mills to calm the obvious discontent.

An angry Reyes Retana argued that the labor delegates represented mill hands. Having signed the agreements, workers had no right to complain. Ramos Pedrueza agreed, adding that if mill hands attacked the contract, they would insult the president of the republic. He held union leaders responsible for controlling workers. Sánchez Gavito claimed that the immediate application of the contract was a question of "social discipline." He demanded the jailing of workers who continued to strike.[134]

The strikers were not jailed, however, so that workers won the next round.

Labor's Response

During the last days of the convention, conflict in the mills continued unabated. Río Blanco strikers refused management's offer of a settlement and continued the work stoppage.[135] Convention delegates received *avisos de descontento* from workers in Tlaxcala, Puebla, Orizaba, and other areas.[136] In Atlixco, workers at El León reported that the government had arrested two union leaders and tried to prevent them from holding union meetings.[137] In Orizaba, administrators posted lists on the factory doors of fired workers: a hundred in Nogales, another hundred in Cocolapam, more in Río Blanco.[138]

The Convention ended on August 1, a day in which Mexico City newspapers headlined "It is rumored that a general strike will break out in the Republic," provoked by "the unhappiness of workers."[139] This was accompanied by threats from union leaders in the textile zones to strike in defense of their new unions. Labor's Central Committee simply declared that "we are not responsible for what might happen in the factories."[140]

There were three problems. First, the new work rules overwhelmingly favored owners, and no amount of rhetoric could make workers see otherwise. Second, while the wage scale raised wages in some mills, it lowered them in others. Third, workers knew that their gains had come from the earlier general strike, not from a meeting in Mexico City. If they wanted more, it was obvious that confrontation yielded more than negotiation.

In high-wage zones like Orizaba, where the *Tarifa* reduced some wages, local leaders urged noncooperation.[141] Unionists telegrammed the Central Committee to advise that local authorities were arresting union leaders.[142] Santa Rosa workers walked out "for not agreeing with the new work rules."[143] In other areas, workers simply tore them off the doors of the mills.[144] There were reports that mill hands in La Trinidad (Tlaxcala), Santiago (Puebla), and San Agustín (Atlixco) followed the example of Santa Rosa workers, destroying the new reglamento when it was posted and going out on strike.[145]

On August 4, some 2,000 mill hands from twenty-three textile factories met in the Teatro Variedades in Puebla to learn about the new contract. Shouts of "death to the industrialists" filled the hall as sentiment turned against the convention.[146] Workers complained about several clauses, including those on cleaning the machines, the number of looms assigned to each worker, distracting fellow workers, and the requirement to work the whole week. The next day, Puebla and Orizaba awoke to another general strike by textile workers. Reported one newspaper, "The question of the approved wage scale has caused such discontent, that in Orizaba and Puebla the strike is almost complete."[147] Mill owners called on the government to increase the military presence around the factories.[148] Meanwhile, members of the Labor Central Committee fanned out in Puebla and Atlixco to defend the agreements. Otilio Venzo, a Puebla factory worker, Juan Olivares, Luis Flores, Rafael Silva, and Alfonso Reséndiz arrived from the capital to calm the mill hands. They argued that the contract was fair; it was only the work of two agitators that led to the strike.[149]

Unfortunately, the problem was not two agitators but rather the general sentiment of rank and file. Puebla Jefe Político Carlos Aldeco, meeting with seven hundred strikers, promised to help with "just petitions" but threatened to punish those responsible for "instigating disorder." He repeated the official line, that mill hands needed to support the new agreements because they emanated from the convention, were sanctioned by government, and provided for wages higher than those in English factories. The mill hands angrily expressed their opposition, in particular to fines and to arbitration commissions which would not work because the owners could fire workers on the commissions who voted against them. When Aldeco asked how they would like to modify the rule, they shouted back "the only change is to erase it." As they began to shout, Aldeco lost control of the meeting and threatened to call the police.[150] The frightened Puebla owners posted announcements in the factories proclaiming the suspension of the new contract, "because of the disagreements."[151]

In Orizaba, the Jefe Político initially took the hard line, arresting five women and two men for throwing bits of tortilla and bread at scabs entering the striking factories. Meanwhile, the subdirector of the Labor Department and various members of the CCOR set out for the town to convince the workers to return to work.[152] Between August 7 and August 8 the strike was lifted, but only after milhands won a victory. On August 7, strikers told the industrialists and political chiefs in both Puebla and Orizaba that they would lift the strike only on the condition that the new wage scale be posted in all the factories and applied within five weeks, but that the work rules be suspended or postponed. When owners and government agreed to these conditions on August 8, the factories returned to work. Afterwards, some factory administrators tried to fire strike leaders, often leading to new strikes followed by readmitting agitators to the mills. Thus, despite the settlement, one newspaper noted that "there subsists an undercurrent of resentment."[153]

In retrospect, the work of Mexico's first official tripartite convention was both a great victory and nothing at all. To get two owners' groups, with some of the wealthiest and most powerful men in Mexico, labor leaders, and government officials to draft a virtual collective contract and begin the institutionalization of the labor relations system was a radical and major achievement. From that day forward, Mexico would march along the path set by the pioneers at the 1912 Convención de Industriales. Present day collective contracts in Mexico date back to those early agreements.

On the other hand, it was simply too little and too late. The world had changed much more than the industrialists realized. Owners were accustomed to running the mills through a system *"de tipo militar."*[154] Ownership made all the decisions. The owners "in the majority cling to their unchallenged authority in the factory."[155] Unfortunately, their unquestioned authority in the factory collapsed between November 1910 and July 1912. In July 1911, Río Blanco labor leader Pánfilo Méndez demanded that the mayor "sacrifice part of his power" in order to let "the people" select the police chief.[156] As the old order collapsed around him, the new mayor of Río Blanco did just that.[157] Just a month earlier, the previous mayor cried that workers "were insubordinate, lacking in respect for the head authority invested in me, and threatening me."[158] "*Se me insubordinó!*" The insubordinates from the laboring classes had acquired the power to name police chiefs, challenge mill owners, and threaten municipal presidents.

The owners were not fools. They agreed that the real problem was this insubordination, expressing itself as a challenge to traditional authority. Factory owner Alberto Sánchez Vallejo blamed it on the successful general strike. "We understand that it is a consequence of their recent triumph [1911 strike], but it is the case that they no longer accept either comment or punishment of any form."[159] He claimed that workers drank too much, came to work inebriated (understandable if they drank too much), then would sneak into the bathrooms to smoke, chat, and nap. They came to work late and left early. They didn't clean the machines, instead often falling asleep on top of them. They wasted material. Whenever they felt like it, without any more restraint than a child, they threw a party, the infamous fiestas that made workers leave the factory despite being in the middle of an important task.[160] "*Hacen todas cuantas fiestas se les ocurre.*"[161] In the old days, owners punished behavior like this. Since the 1911 general strike, the owners felt there was little they could do. Fellow owner José González Soto called them "these foolish workers."[162]

While the early revolution may not have changed feelings about class, it changed feelings about how those in the underclass could behave with those in the upper class. The ambiance of revolution legitimated a workers' challenge to the authority and dignity of owners. That is what owners understood as insubordination. When agitators like José Otáñez insulted owners, they found overwhelming support among the rank and file.[163] Between 1910 and 1912, workers came to accept not only the legitimacy but also the utility of such challenges. In March 1912, a worker named Emiliano in the San Bruno mill unintentionally broke a tool. The factory followed its normal practice of taking a weekly deduction from his wage until he paid for "the small damage." The other mill hands considered the fine "very severe and little justified," and threatened to "make common cause with the fined worker, abandoning work next Monday." This was a collective challenge to the authority of owners to enforce labor discipline. The federal government, under the pressure of a widening revolution, argued that "in the current circumstances, any kind of strike could lead to serious disruptions of public order, the government is very interested to avoid paralyzing the factories, and for this reason recommends that you try by all possible means to avoid the developing conflict."[164] Despite the warning, the administrator fired Emiliano for challenging his authority. It was necessary, he said, "to avoid that these things repeat themselves with frequency given that workers are so distracted in their tasks, attending more to gossip than to their machines . . . You'll understand that this matter can have no other solution; otherwise it would be surrendering the authority principle (*perdiendo el principio de autoridad*), these cases repeat themselves endlessly, and for insignificant things."[165]

That same month, mill hands in El Carmen carried out a short wildcat strike in order to challenge an old custom. Traditionally, weavers carried their new cloth to the office for weighing, which would determine their pay. Suddenly, the manager complained, they walked out in order to force the mill to hire workers to carry the cloth for them. They "refuse to work without just cause."[166] In July 1912, the Mayorazgo mill hands also carried out a strike against "established work customs."[167] Challenges to "established customs" were the most profound challenges of all because the strongest authority is that not thought about. If not spoken, it was not questioned and therefore not questionable. During revolution, however, everything was questioned. As the owners of El León wrote, "For the last two years, workers have become insubordinate, there is no way to make them work in an ordered fashion, they no longer respect their superiors . . . these workers are very insolent."[168] Once again, the accusation of insolence.

In January 1910, cotton textile workers saw themselves as an exploited class, but there was little they could do about it. In November, Madero's call to revolution quickly toppled the old regime. Sensing weakness, mill hands organized new trade unions or reenergized old ones and carried out the first industrywide general strike in Mexican history. Between December 1911 and January 1912, they won new power. Without the old repression, textile owners conceded higher wages and lower hours of work.

Fearful, government and owners organized a convention. The meeting ratified the first nationwide wage and work rules in Mexican history. Rebellious textile workers had forced the country to begin to institutionalize the Mexican labor relations system.

Fortified as much by the weakness of their enemies as by their own strength, the rank and file continued with strike movements through the weeks of the Convention and afterwards. The owners recognized the problem. The collapse of the Díaz government provided an opening to workers, and the success of their early movement changed their mentality. They became "insolent," which meant they came to believe in challenging authority. Since authority at work rested on the traditional rights of private property to command, the workers' revolution quickly developed into a threat to authority at work and to the rights of property holders. It was that continuing challenge that owners saw as the insolence of workers, a lower class who refused to act like a lower class should. In the absence of effective repression by the central government, workers won the opening of their revolution.

CHAPTER 5

Challenging Authority, 1912–1916

AUTHORITY IS THE CENTRAL QUESTION in every social revolution. The insolence of workers during the early years made authority a problem in textile factories. This chapter examines the degree to which their challenge to authority fundamentally altered the social relations of work in Mexico, private property rights, and the labor regime. Second, it explores how challenging authority changed during the course of revolution. As will be seen, authority relations in the factory by the end of the revolution were not like those at the beginning.

Acceptance of authority and hierarchy is ultimately an embedded social process. At work, it is a function of the labor regime, the "rules, habits, norms, conventions and values" of work and the social relations of work.[1] Within these habits and norms, none are more important than the relationships of command and obedience.

While the general premise of a market economy is the authority of capital, the legitimacy of that authority, indeed its specific reach, can vary greatly from one society to the next. Historically, private property rights are not only a legal but also a social concept, particularly though not exclusively with regard to command and obedience. Although property rights normally include the right to command at work, the constraints of such commands have varied greatly over time and from one society to another. Whatever the constraints, when the legitimacy of command breaks down, the old labor regime collapses because capital can no longer rule labor. The formal mechanisms that enforce authority at work are normally not sufficient to make the work place function without the legitimacy of that authority.

The starting point for the workers' revolution was the Porfirian labor regime. During the long prerevolutionary industrialization, Mexico constructed a labor regime that relied on government support of industrialists and repression of workers, on the modernity and legitimacy of the new factories, and on the transfer of the hegemony of hacienda workplace relations to the mills. Whenever a growing working class attempted to challenge the new social relations of the factory, effective repression stopped them.

Through 1909 and 1910, diverse strike movements hit the mills. What changed in 1911 was the collapse of the Diaz government, so the general strike in December confronted a weak rather than a strong state. When the mill owners could no longer count on effective government support, mill hands were able to quickly build strong unions, force a change in wages and hours of work, and begin to challenge the workplace authority of owners.

The challenge to authority that characterized the beginning of the workers' revolution deepened and evolved in subsequent years. This chapter follows the process from the close of the textile convention in July 1912, to the opening of the Constitutional Convention in December, 1916. First, it describes the challenge to authority in the mills immediately following the convention, 1912–13. It will show how a laborite victory in the general strike, combined with the absence of strong government in the sudden change from Madero to Huerta, provided workers with incentives to pursue a more radical revolution in the workplace. As their revolution became more radical, it forced aspiring elites to institutionalize the labor relations system, initially through military decrees. Between 1913 and 1916, regional military commanders promulgated military labor decrees that attempted to stop the labor revolt through law while ceding to workers their demands. These decrees later became the basis of Article 123 of the 1917 Constitution, demonstrating that Mexico's labor regime did not emanate from above, from the state, but rather from below, from the workers' challenge to authority. It is important to understand the military decrees as part of the process of the workers' revolution, in which labor law emerged from the battleground of the mills and subsequent if indirect negotiations between a rebellious working class, angry owners, and contentious and aspiring elites, in the midst of revolution and civil war.

The third section explores the process through which the challenge to authority evolved in the course of revolution. Before the revolution, it was daring for workers to unionize or strike. During the early revolution, mill hands added new workplace and shop floor demands. The military decrees attempted but failed to silence workers over these issues. As a consequence, between 1913 and 1916, mill hands increasingly fought for control over the workplace. In many cases, this meant challenging foremen or preventing firings. In others, it meant getting rid of supervisors or having the union control who worked in the mill. It some cases, mill hands demanded strengthened rights for unions and more respect for workers. These demands, often backed by wildcat strikes or other labor actions, and the inability of owners to fire workers or repress unions, gave workers and their organizations greater power in the workplace. In this way, challenging authority during the later revolution was also a challenge to traditional property rights.

In tracing these events, this chapter shows how a broad but thin workers' movement evolved into a true social revolution in the workplace, which is to say, a generalized challenge to the authority of owners at work. The initial push was broad because the 1912 general strike essentially shut down the entire Mexican cotton textile industry. It was thin because it sought traditional laborite goals, wages and hours of work. That is why the owners could easily cede these points. Even during the early revolution, however, workplace issues came to the fore. As the revolution progressed, as no strong state came to the fore, as workers gained increasing self-confidence, mill hands came to believe that their labor organizations could replace capital as the dominant power at work. How far could workers take this challenge to the authority of capital?

Challenging Authority 1912–1913

Before the revolution, textile workers organized unions and carried out strikes. Despite their efforts, the determined opposition of owners, a strong state, and an effective repressive apparatus kept them in their place. None of their struggles produced permanent gains, legal recognition of unions, or institutional limits to the power of capital on the shop floor.

Suddenly, Madero's revolution seemed to change everything. No sooner did Madero issue his challenge to Díaz than textile workers organized unions throughout the industry. When Díaz fell and León de la Barra assumed the presidency, mill hands organized the 1911 general strike. Although owners vociferously opposed labor organization, successively weak presidents—Díaz in his last days, an interim León de la Barra, and an unsure Madero—were unable to repress workers. Given this context, labor organization and activism produced the higher wages and lower hours of work that were ratified in the July 1912, convention. While the owners felt that this was revolution and socialism, for many workers it was only the beginning. The victories of 1911 and 1912 ratified their belief that they could now obtain redress for numerous shop floor grievances. Without much organized ideology, mill hands expanded their struggle in a very empirical fashion, as each successful action taught them how to move to the next successful action.

Prior to the revolution, the shop floor was a space of organized hierarchy. It had strict rules whose enforcement rested on the authority of supervisors, which in turn rested on the authority of owners and the ability of the state to support the rule of private property at work. With the state collapsing, and emboldened mill hands pushing forward, what began as a challenge to supervisors quickly led workers to confront the authority of capital itself.

Not all workers challenged authority, nor did all workers agree on fighting the owners. Often their challenge to capital lacked ideological consistency. Nonetheless, enough workers challenged authority enough times with enough success to ultimately smash the old labor regime and with it the prerevolutionary authority of owners. Cotton textile workers entered their revolution like other Mexicans, without a plan. Driven by immediate grievances, the process of revolt took them to new and unexpected territory. It was revolution as process rather than ideology.

The owners reacted by trying to preserve as much of their wealth, power, and privilege as they could, and "The majority clung to their [previously] <u>unchallenged authority</u> in the factory in their dealings with workers."[2] Most thought that the 1912 agreements would return peace to the mills, otherwise they would not have agreed to limiting their authority. The workers, however, used the accords as a springboard to other goals, possibly because the rules of the new institutionalization were unclear. Not surprisingly, worker activism angered the owners, one of whom complained that his mill hands just "demand and demand rights and more rights, concessions and more concessions." He concluded by threatening that "sooner or later we will end this by putting a yoke around their neck, centuries will pass before they can throw it off, because Capital is so brave and strong when it finds no other outlet."[3]

The 1912 agreements did not restore the old authority of capital. Instead, they opened the door to further conflict because mill hands saw them as negotiated documents that resulted from a successful general strike and with some unclear rules. A lack of clarity meant further negotiations, which workers knew came from strikes and labor actions. There were numerous complaints from mill hands to Ramos Pedrueza that the mills either did not fulfill the agreements, or did so in ways that prejudiced workers or unions. Many mill hands, like those in the Santiago factory in January 1913, saw it as a question of "another interpretation."[4] Some of these interpretations could be quite complex because the mills (and workers) needed to calculate piecework rates by machine, parts of machines (spindles, etc.), number of workers assigned to the machine, type of cloth, quality of the cloth, and other factors. Innumerable complaints and conflicts were inevitable.[5]

In late December 1912, the administrator of La Trinidad attempted to implement work rules "in which it is strictly prohibited that any worker bring any food, fruit or drinks into the mill, as well as smoking, and any worker violating this will be expelled immediately."[6] The *operario* Leandro García replied that it was the factory administrator who "commits these abuses, smoking inside the mill." La Trinidad's mill hands walked out "until they get rid of these rules."[7] When a supervisor "began to laugh at the workers," they added his firing to their demands. A meeting of the administrator, union leaders, and the *Jefe Político* of Apizaco got the workers what they wanted.[8] This would have been inconceivable eighteen months earlier. Workers now challenged and changed work rules and, in effect, fired supervisors. Just six months after the Convention, and in one of Mexico's largest mills, workers had gone far beyond the wage and hours demands of the early revolution. When Manuel Ortega Elorza visited La Trinidad in June 1913, he commented that "the workers of this factory frequently violate the Reglamento, missing work without reason."[9]

The conflict with a supervisor was not surprising. In many workplaces, the most common conflicts are between workers and their immediate bosses because the latter are in direct contact with the former. For most workers, supervisors are the face of capital in the factory. As such, they quickly became a common target of worker anger. In November 1913, Mirafuentes workers walked out, demanding "firing of the boss."[10] The owners resisted what they labeled "complaints without basis," because they could not retain their authority if they let the workers choose *maestros*.[11] In this case, they defeated the strike within a few days, fired seven young men, and ordered a return to work "under the same conditions."[12] So it went with the workers' rebellion in the years after the Convention, some victories, some defeats, but increasingly challenging the traditional authority of capital.

The Tarifa de Salarios also opened the door to conflict because it raised wages in some mills while lowering them in others and did not match expectations in still others. In the waning months of 1912 and the early months of 1913, workers in many mills engaged in strikes and walkouts "for not having implemented the new Rules and Wage Scale."[13] When the Miraflores workers walked out, it "caused the justified disgust of [owner] Señor Don Hipolito Gerard who ordered the closing of the mill."[14]

The unsettled situation led to a new strike movement spreading throughout the industry. In December 1912, Ramos Pedrueza ordered labor inspector Miguel Casas to Jalisco "with the goal of solving the difficulties between workers and owners caused by the new wage scale." Casas believed that local workers in Jalisco were "submissive and obedient,"[15] but in the nearby Río Blanco (Jalisco) mill, workers walked out, demanding that the factory implement the new pay scale. To get them back to work, the factory had to give in to their demands. Even mill hands considered docile before the revolution had now become threatening.

In Atlixco, the conflict over the Tarifa led to permanent conflict throughout 1913, with walkouts in Metepec (January and September 1913), El Léon (June 1913), El Carmen (January 1913), and San Agustín (January 1913). The mill hands mostly won these disputes because, as the administrator of San Agustín noted, "in order to avoid a continuation of the strike, I agreed to pay the discontented ones."[16] Owners and government officials were dismayed by the boldness of workers and the inability of the new agreements to slow worker radicalism. On January 2, Ramos Pedrueza issued a stern warning to the workers to "stop the useless and prejudicial strikes," recommending they abandon the "clumsy and destructive custom of wanting to solve all conflicts by means of strikes."[17] Meanwhile, persistent strikes in Mexico's largest factory industry contributed to a growing belief that Madero had no effective strategy to pacify the country. When in the following month Huerta deposed and assassinated the president, some industrialists claimed that "in a short time the work rules and wage scale [agreements] will disappear and we will return to the old system."[18]

However, there was no return to the old system, not under Huerta and not later. General and President Huerta confronted an extraordinarily difficult situation. With Zapata in rebellion in Morelos and Carranza organizing resistance in the north, his effective rule was limited to the cities, where textile workers were strongest. He could ill afford a frontal assault on them. When labor leaders informed Huerta's Labor Department that the owners wanted to get rid of the new agreements, the department tried to "eliminate in the workers the false impression that some industrialists have tried to spread."[19] Huerta, like Madero, lacked sufficient strength to reimpose owner control.

On the other hand, local authorities and industrialists did their best to fire or jail labor activists. In April 1913, factories in Toluca fired union activists "for spreading the seeds of discord among peaceful workers."[20] In July 1913, Río Blanco used its influence with local government to have union leader and "instigator" José Natividad Díaz jailed for 30 days.[21] La Trinidad fired Gabino Sánchez, the head of its union. Sánchez moved to Puebla to look for work but the mills refused to hire him. He eventually found a job in Santa Elena, where after months of begging, he convinced the administrator to hire him "with the condition of not getting involved with these [union] affairs," adding that "they watch my every movement inside the mill."[22] Activist Natividad Díaz was more successful, reappearing later as union leader in other mills, as militant as ever.

The women in the mills participated in the fight against authority. In March 1912, the women in La Corona (Puebla) went on strike "against bad treatment."[23] In August, the women in El Hércules demanded the removal of a supervisor, complaining that Octavio Regalado "is so abusive that he spends most of the day walking through the department where we work, mistreating us and saying things that we as women cannot refer to."[24] The male union leader, José Pérez, then added a demand to fire another supervisor for his "arbitrariness" with the women.[25]

Thus the year following the convention was filled with as much if not more conflict than the year before. Workers won an unusual number of their strikes, which began to include such demands as firing unpopular supervisors. Women and workers outside the core textile zone became radicalized. The new industrywide agreements seemed to empower workers. In El Batan, mill hands walked out in solidarity with a worker who was fired for drinking on the job.[26] In La Constancia, they celebrated a holiday without advising the mill, so the owners temporarily shut it down "in order to punish its workers."[27] If there was no peace in the country, there certainly was no peace in the mills.

The Military Decrees

After Huerta assassinated President Madero and Vice President José Pino Suárez, on February 22, 1913, revolutionary violence exploded across the country. On February 23, the Maderista governor of Coahuila, Venustiano Carranza, announced his opposition to the new president, whom he labeled a usurper. In March, the old Maderista revolutionary, Pancho Villa, returned to Mexico via El Paso to join Mexico's second revolution. In May 1914, Huerta submitted a law to extend the textile tax, the enforcement mechanism for the 1912 textile agreements. It was one of the last acts of his government before he fled to Europe in July. As violence increased throughout the country, many foreign nationals also fled to Europe, including an important contingent of textile technicians.[28]

Soon after, the victorious revolutionaries, Carranza and Obregón on one side, Villa and Zapata on the other, fell to fighting among themselves, launching the country's third revolution. It lasted until late 1916 and early 1917, when the drafting of the new Constitution signaled the effective victory of Carranza's Constitutionalists. During this period, the textile economy suffered, and the civilian population in the textile zones experienced the horrors of widespread killing and destruction. Despite the difficulties, however, most mills continued to function and most operarios continued to work. The immediate consequence of the collapse of the Huerta government was a short-lived strike movement, "as a consequence of the current political change."[29] Shortly thereafter, in many parts of the country true authority fell to local military commanders who became effective government. Given the unsettled conditions of social revolution and civil war, the commanders needed as much popular support as they could get.

In the textile areas, the generals desired peace in the factories, as had before them Díaz, León de la Barra, Madero, and Huerta. Mostly Constitutionalists, they responded to continuing labor conflict and labor violence by issuing military labor decrees for the regions under their control: Aguascalientes (August 1914), Chiapas (October 1914), the Federal District (September 1914), Guanajuato (December 1914), Jalisco (October 1914, December 1915), Michoacán (October 1914), Puebla (September 1914, September 1915, December 1915), Tabasco (June 1913, September 1914), Tlaxcala (September 1914), Veracruz (October 1914, January 1916), and the Yucatán (September 1914).[30] Colonels and generals—"El C. Coronel Abel B. Serrato, Jefe Político del Distrito y

Comandante Militar de la Plaza," or "Luis F. Domínguez, Gobernador Militar del Estado de Tabasco"—rather than lawyers and politicians initiated this second institutionalization of a labor relations system.

Because the oldest and most extensive textile conflicts were in Puebla and Veracruz, military commanders in those states issued the earliest, most comprehensive, and most radical decrees. On September 2, 1914, the military governor of Puebla, Pablo González, "*General en Jefe del Cuerpo de ejército del Nordeste*," declared "the misery of the lower working and peasant classes, exploited during a long time by the bourgeoisie who has accumulated wealth at the expense of the suffering and excessive labor of the class' misery, and taking into consideration that one of the reforms of the Revolution is to elevate, as much as possible, the economic conditions of this great majority."[31] He issued a seven-article decree that established minimum wages and maximum hours of work, neither of which existed by law. The decree included some educational benefits but ignored the most conflictive issue in the workplace, unions. It did add a system of state inspectors to "supervise its exact compliance."

The decree set off "some difficulties."[32] Local textile owners were particularly angry about the eight-hour day because it contradicted the ten hours of the 1912 contract. They claimed that it placed them at a competitive disadvantage with mills that operated under the national agreement. After a meeting with labor leaders and González, they obtained a temporary nine-hour workday in Puebla and Tlaxcala textile mills.[33] The CIM then asked Constitutionalist leader Venustiano Carranza to establish federal law in order to avoid the competitive disadvantages of state labor law.[34]

A year later on September 22, 1915, "*El C. Coronel Luis G. Cervantes, Gobernador*" decreed the establishment of a state "Oficina Técnica del Trabajo" with power to "intervene as arbitrator in problems between owners and workers."[35] On December 4, 1915, he signed a further decree that gave state recognition, and thus legalization, to unions. The decree also established a *Junta de Vigilancia de Patrones y Trabajadores*, which gave unions the exclusive right to bring complaints to the Labor Board.[36] In little over a year, military decrees in Puebla not only legalized and recognized unions but virtually mandated them through their exclusive authority to bring complaints to the state labor boards. They also mandated minimum wages and maximum hours of work and provided some benefits, earning the hostility of owners.

Marjorie Ruth Clark commented that neighboring Veracruz was "the scene of more disturbances between labor and capital than any other section of Mexico."[37] On October 4, 1914, "*El Gobernador acciden-tal, Coronel Manuel Pérez Romero*," issued a decree that established the Sunday day of rest.[38] The "accidental governor" did not remain long in office. Two weeks later on October 19, 1914, Cándido Aguilar, "Gober-nador y Comandante Militar," issued Decreto Número 11, which estab-lished a minimum wage, overtime pay, and medical and educational benefits. It implicitly recognized unions by allowing them to bring complaints to local governments.[39] Aguilar continued his reforming ways in January 1916, with Decreto Numero 15,[40] which provided a legal definition and state registration of labor unions. It also gave unions the right to sign contracts and set fines for businesses that refused to recognze them. When asked about the labor problem in Veracruz in late January, Aguilar responded "I consider it almost solved with the recent union law I just signed."[41]

Thus between September 1914 and January 1916, military command-ers in the core textile zone, Puebla and Veracruz, created a new institu-tionalized labor relations system that included union recognition, state-mandated minimum wages and maximum hours of work, and state-mandated medical and educational benefits for workers. The new system conferred legal authority on state governments to arbitrate labor conflict, to enforce labor law, to protect labor organizations, and to set working conditions by political rather than economic means. Decree Number 11 in Veracruz became the most influential: it established state authority in labor affairs through the creation of state labor offices and inspectors and the implementation of wage and labor standards with universal application. It made minimum wages, overtime pay, maxi-mum hours of work, union recognition, and benefits a matter of state policy. Such decrees appeared in other states, as seen in Table 5.1.

In addition to the state decrees, Venustiano Carranza, "*Primer Jefe del Ejército Constitucional*," issued two military decrees that sought to "avoid the current conflict between . . . industrialists and workers."[42] In December 1914, he mandated an extension of the 1912 textile agree-ments. He also increased wages between 20 percent and 30 percent and reduced the workday from ten to nine hours. The decree virtually le-galized and encouraged unions, entrusting textile workers to "elect unions (*juntas directivas*) in each factory to represent their demands in each case of a violation of the Uniform Minimum Wage Scale and Work Rules."[43] On March 22, Carranza mandated another wage hike (35–40%) for textile workers.[44] In between the two textile decrees, on January 6, 1915, he issued an agrarian law to restore peasant lands and on January 29 a decree to federalize labor law.[45]

From these decrees, it is clear that Constitutionalist military commanders, including Carranza, believed that they could not pacify the country without pacifying the mills. As revolutionaries, they could not repress the underclass without losing legitimacy, so they continued to follow the path of an institutionalized labor relations system. They believed they could control workers and unions by placing their activities under some sort of state supervision. On April 12, 1915, Carranza's minister of government, Rafael Zubáran Capmany, submitted a *"Proyecto de Ley sobre Contrato de Trabajo."*[46] The proposal argued that:

> the relationships between Capital and Labor have been taking since then [Constitution of 1857] a character of hostility that before they didn't have, and that have been exacerbated to the degree that the development of the capitalist regime, whose clearest manifestations have been *maquinismo* and the concentration of industry in large factories, has made more frequent occasions of conflict between these two factors of production, whose harmonious participation constitutes today a very distant ideal.[47]

The proposed law, forty-eight pages, seven chapters, and one hundred articles, defined the labor contract, the rights and obligations of workers and owners, the length of the working day, minimum wages, work rules, legal firings, the legal rights of unions, and labor institutions for federal, state, and municipal governments. It was a radical break with the Porfirian past. Unlike the 1857 Constitution, it did not leave labor affairs to the market. Instead, it provided government definition and regulation of contractual matters, government enforcement of clearly defined work rules and behaviors, government standards for compensation and benefits, and federal and local government enforcement. It prohibited the worst abuses of the past, such as the company stores and employment blacklists. It provided protections for unions, unionized workers, collective contracts, and jobs. It made it legally difficult to fire workers or for employers to block workers' associations. Zubáran Capmany's proposal also tried to advance the status of women. He noted that Mexican common law considered adult married women "incapable of contracting, obligating, or litigating without the consent or authorization of the husband."[48] In contrast, his proposal allowed adult married women to sign work contracts on their own, because of *"the scarce budget of the worker household . . . it was necessary to liberate the wives."*[49] These provisions placed the state directly in the middle of the work world.

History has mostly ignored Zubáran Capmany's *Proyecto de Ley* because it never became law. Within two years a new and revolutionary constitution rendered it obsolete and irrelevant. Written in Veracruz

TABLE 5.1
The Military Decrees

Decree	Authority/author	Legal work day	Minimum and other wages	Benefits	Unions	Government offices
Tabasco[i] June, 1913	Agustín Valdes, Gobernador Interino			Arbitration Boards		
Aguascalientes, August 1914	Alberto Fuentes, Gobernador y Comandante Militar	9 hours		Sundays		
Distrito Federal September 1914	General Heriberto Jara	9 hours		Sundays, holidays		
Puebla, September 1914	General Pablo González	8 hours	.80 pesos	schools		state labor inspectors
Tabasco[ii] September 1914; "Decreto Relativo al Proletariado Rural"	Gobernador Militar Luis Domínguez	8 hours	1 peso			state labor inspectors
Chiapas, October 1914, "Ley de Obreros"		10 hours	1 peso, industry; 50% overtime	holidays; schools; medical costs, 50% sick pay; housing		
Michoacán, October 1914	General Gertrudis Sánchez	9 hours	.75			

Decree	Authority/author	Legal work day	Minimum and other wages	Benefits	Unions	Government offices
Veracruz, October 4, 1914	Coronel Manuel Pérez Romero, Gobernador Accidental			Sundays		
Veracruz October 19, 1914	General Cándido Aguilar	9 hours	1 peso; 100% overtime	Sundays, holidays; schools; medical costs; 100% sick pay	tacit recognition	state labor inspectors; municipal labor authority
Guanajuato December 1914	Coronel Abel B. Serrato			Sundays		
Industria Textil, December 1914	Venustiano Carranza, Primer Jefe del Ejército Constitucional	9 hours	mandated pay raise		Required unions	
Industria Textil,[iii] March 1915	Venustiano Carranza, Primer Jefe del Ejército		mandated pay increase			
Puebla September 1915	Coronel Luis Cervantes					state labor office
Puebla December 1915	Coronel Luis Cervantes				formal state labor recognition	court
Veracruz January 1916	General Cándido Aguilar				formal recognition	

i Rural workers.
ii Rural workers.
iii Textile workers.

when Carranza did not control the country, many of its ideas did not resurface until the 1931 Ley Federal del Trabajo. However, its importance is not in what the proposal accomplished but what it signified. First, it suggests the link between labor conflict in textiles, the Constitutionalist military decrees, and Article 123 of the 1917 Constitution. Militant workers had forced military commanders to recognize unions and reduce the length of the workday, gains ratified by Carranza's decrees in late 1914 and early 1915. Workers had moved beyond wage and hour demands in order to fight for control over hiring and firing in the workplace as well as other measures that would give unions real power at work. Radical and successful workers forced aspiring elites to rethink the Mexican labor question. As a result, aspiring elites came to understand that there could be no pacification of Mexico without a new and institutionalized labor relations system that made real concessions to workers. Zubáran Capmany's *Proyecto de Ley* demonstrates that the workers' revolution of the 1910s drove legal change, not politicians worried about losing elections in the 1920s.[50]

The Continuing Challenge to Authority

Although authorities expected the decrees to pacify the conflictive workers, they did not. First, because trade unions led the workers' revolution and the decrees recognized and protected them, new law gave workers and their organizations an increased capacity to challenge owners. Second, workers took the decrees as rewards for their militancy, increasing their confidence in their ability to gain more control over the workplace. Third, the creation of new institutions also created many new undefined areas in which power between owners and workers could only be fixed through further struggle. Thus the decrees did not quash the rebellion.

In Orizaba, the workers defiantly proclaimed "And once and for all, let the industrialists know that the period of tyrannies has passed."[51] In December 1914, mill hands throughout the region carried out a series of strikes in order to gain enforcement of the new decree. When nobody lost his or her job for walking out, confidence in militancy increased.[52] The Cerritos mill hands wrote to Marcos López Jiménez, the new head of the Labor Office, to complain that the mill was not fulfilling the terms of the 1912 contract.[53] López Jiménez traveled to Orizaba to ask the mill's workers whether they had a union representative. On hearing that they did not, he helped them elect union officers, tantamount to organizing a union. The very next month the newly elected officials, led by José Natividad Díaz, demanded the implementation of Cándido Aguilar's decree, which would reduce the workday to nine hours, increase pay 20 percent, and allow for official recognition of the union. The company immediately granted a 10 percent pay hike, agreeing to satisfy the other demands if more factories followed suit.[54] Thus in Orizaba, unionization brought the labor decree, and the labor decree brought unionization.

In neighboring Puebla, Pablo González's decree dropped the legal workday to eight hours, which immediately caused "incidents between owners and workers in the cotton textile industry."[55] The owners fought the measure, which could lead to "the death of the industry in the region."[56] The old Comité Central Permanente de Obreros de Hilados y Tejidos de la República led the fight for the decree, arguing that "the industrialists have exploited [the workers] for a long time, restricting their liberty and violating their rights with impunity."[57] Arduous negotiations between the Labor Office, General González, and the CIM brought a temporary agreement to implement the nine-hour day, still a reduction from the ten hours mandated by the 1910 contract, and to send separate worker and owner commissions to speak with Carranza about a federal labor law.[58] As we know, Carranza responded by federalizing labor law and decreeing the nine-hour day for textile workers throughout the country.

As mentioned earlier, the military decrees put the state directly in the middle of labor-capital conflict. An example of the new circumstances took place in 1916 when the Santa Rosa union demanded 100 percent overtime pay for Sunday and holiday work. Owner Camilo Maure rejected the proposal, and the new Veracruz Labor Board tried to arbitrate, offering the workers 75 percent, which union leaders declined.[59] At the next union meeting, workers spoke strongly against the company's position. One worker disputed the mill's position that overtime work was inferior to regular work, another argued that the factory administrator was a paper pusher and therefore not competent to judge real workers, and still another said that overtime deprived them of the right to "rest peacefully at home." One union member concluded that it would be better for the mill hands to refuse to work on Sundays if *their* conditions were not met. The workers concluded by voting to refuse overtime work at less than 100 percent.[60] The local labor board ruled against the workers, voting that Sundays and holidays merited only a 75 percent premium, which Maure immediately accepted. The union, however, got the board to agree that workers could not be forced to work overtime; they exercised their rights by simply refusing to enter the factory on Sundays and holidays.[61] In this way, Cándido Aguilar's and Heriberto Jara's labor boards protected workers even if they did not always rule in their favor. The cost, however, was that the state became a determinant in things intimate to workers, like dignity, personal life, and private amusement.

Meanwhile, the decrees in no way ameliorated fights against supervisors, a de facto bid for control of the shop floor. In early November 1914, Daniel Galindo, federal labor inspector, and Tomás Heredia, from the Workers Permanent Commission, traveled to San Martín Texmelucan, site of the San Félix mill. A group of workers had walked out in order to get a hated supervisor fired, accusing him of "mistreatment." The union leader tried to convince the mill hands to not strike, but when they saw the *cabo* "with a haughty and hostile attitude," they walked out again.[62]

In January 1915, the San Lorenzo union invited López Jiménez to visit the mill to witness the "violations of the Reglamento and Tarifa Mínima." Workers complained about the factory administrator and several supervisors."[63] They claimed that the "foreign Director . . . did not listen to the operators," requiring that they speak instead with a supervisor, Joaquín Badiano. They reported that Badiano "treats the workers with insolence and contempt." They accused the administrator of trying to keep unions out of the mill and the supervisor of violating the 1912 contract and requested that both receive "un castigo ejemplar."[64]

Sometime thereafter, Manuel Sánchez Martínez, the new union head, led a workers' commission to see the Nogales municipal president to complain about supervisors mistreating workers.[65] The municipal president then sent a letter to Joseph Taylor, who ran the factory, asking him to get the department heads to improve their behavior.[66] Taylor denied the workers' accusations, substituting his own charges, that the mill hands were no longer willing to "conform to . . . established customs" and that they now refused to clean the machines twice a week.[67]

An angry letter followed in which the workers focused on "el maestro Pascual Pérez," whom they accused of being "an inept despot and abandoned of his obligations . . . if we insist with him, he says that if one goes along, fine, if not, the street is wide."[68] They also accused the supervisor of physically abusing the boys who worked there, insulting the men, and cheating the workers out of their wages.[69] Some days later the conflict escalated when there was either a lockout or a work stoppage in the department of weaving. Federal labor inspector Manuel Díaz met with the Presidente de la Junta Civil, joined by the president and vice president of the union as well as Taylor and Pérez. Sánchez Martínez repeated the accusations against Pérez and insisted on his firing. When Díaz sided with the union, Taylor agreed to fire Pérez, but also Sánchez Martínez, whom he fired on the spot because "this one for no reason will return because he is constantly making accusations and refutations without any basis."[70] The firing was in vain, however, as Díaz used his authority to protect Sánchez Martínez's position in the mill.[71]

Taylor was understandably furious. He fired off a letter to the Labor Office in which he denied the workers' accusations. He clearly understood what was at stake in the fight, arguing that "to give in to the desires of the workers to fire directors [department heads] is to invite disorder and disaster, because without discipline and supporting them outside of any logic or proof, would make it impossible for us to work."[72] Taylor demanded respect for the authority of bosses. Despite his desires, however, foreman Pérez lost his job, union leader Sánchez Martínez kept his, and the factory and the union signed an agreement that gave the mill hands a new maestro with whom they were "very content."[73] The military decrees had opened a world of conflict, had brought the state into labor affairs, and had provided workers the capacity to effectively challenge authority at work, which sometimes meant getting rid of hated bosses.

The federal Labor Department blamed the Orizaba conflict on outsiders, mostly the Casa del Obrero Mundial. It noted that "The Obrero Mundial riles up the majority."[74] The Casa was active in Orizaba during this period. On March 15, 1915, the vice president of the Río Blanco union attributed a walkout to the activities of Casa organizers.[75] Casa speakers told mill hands that the Labor Office was useless, that the 1912 textile contract would not help them, and that for workers to achieve their aims, "they would have to achieve them by bullets and not by laws."[76] However, challenging authority began in Orizaba before the Casa arrived, and continued after it left, so it is not clear that the Casa was indeed the cause of worker unrest.

The real problem was that it had become legitimate for mill hands to challenge authority in everyday work practices such as showing up for work or obeying orders. At San Félix in late 1914, mill hands walked out to get a foreman fired.[77] In early 1915 at Cocolapam, they walked out to hold a union meeting. They had not advised the mill, which then tried but failed to fire them in retaliation.[78] Strengthened, mill hands then began to ignore the commands of supervisors, as when the *Correiteros* "did not obey him since he became the head of the Weaving Department, as *maestro* and substitute for the director."[79] The Labor Office negotiated a truce in this case, making both sides promise to be nice to the other, but it is hard not to see an insincere agreement in which each side was biding time in order to defeat an enemy.

In June 1915, another wave of walkouts hit Orizaba. Although mill hands expressed loyalty to the Constitutionalists, local authorities thought they were "too violent."[80] On June 17, mill hands marched from the Santa Rosa mill, the most distant from the city, toward Orizaba. As they marched through Nogales, San Lorenzo, Mirafuentes, and Río Blanco, workers from the other mills joined them. By the time they reached Orizaba, where mill hands from Cocolapam and seamstresses from La Suiza joined the contingent, there was a prostrike demonstration of "more than 10,000 workers."[81]

The owners fought back with blacklists. By 1915, the Orizaba mills had instituted a blacklist of agitators and other workers who:

1. became union leaders,
2. signed labor petitions,
3. demanded justice,
4. did not arrive to work early,
5. stayed too long in the bathroom, or
6. were simply suspected of standing idly in front of the machines.

The factory administrators circulated the lists among themselves, and the blacklisted mill hands found it impossible to work in the factories.[82] Although workers on the list responded by applying for jobs under false names, they were immediately fired if discovered.[83] Even so, the blacklists mostly did not work, as the ongoing walkouts, strikes, marches, and protests suggest.

While textile workers were most active in Orizaba and Puebla, there were also waves of strike activity in Mexico City in 1915 and 1916. In July and August 1916 the Casa del Obrero Mundial led a general strike of workers of many industries. With great violence, Carranza crushed the strike and virtually destroyed the Casa. Although some historians interpreted this as the beginning of a decline in the workers' revolution, the Casa strike had surprisingly little impact on textile workers.[84] In Mexico City, textile workers won a 200 percent wage increase. There were similar strikes and outcomes in Tepeji del Rio and Orizaba. Puebla mill hands won a 300 percent wage hike by threatening a strike.[85] There, a new Confederación Nacional del Trabajo organized a meeting in November 1916 to press for wage increases, which the governor supported.[86] At the same time, Veracruz workers sent a commission to Mexico City to get the federal government to support gold equivalents for money wages.[87] Cándido Aguilar, now the federal minister of foreign relations, backed the commission, as did Heriberto Jara.[88] A few days later, the mill hands obtained an important victory when the federal government in Mexico City authorized state governors to regulate a new wage system based on gold equivalents.[89] In Puebla, Veracruz, and Mexico City, the heart of the textile industry, labor militancy continued unabated, with important victories for workers.

On November 18, 1916, the military commander and governor of Mexico City, General César López de Lara issued regulations to protect workers' salaries. Three days later, the city's newspapers announced that *"El Congreso Constituyente Inicia Sus Labores,"* as the Carrancistas met to draft a new constitution.[90] By now, the workers' revolution guaranteed that the Congreso Constituyente would address the labor problem. New elites needed new law to match the social reality of Mexican factories, where authority had become contested terrain.

Thus, labor militance continued unabated between 1913 and 1916. Mill hands continued to build their unions and demand higher wages. They increasingly employed wildcat strikes and walkouts to get rid of foremen and supervisors, often with success. They also challenged the mills on such things as cleaning the machines or working overtime. The set of such challenges inevitably weakened the authority of factory administrators to run the mills. Furthermore, the constant strikes over interpretations of the 1912 agreements made all work rules subject to negotiation between owners and unions, in practice if not in law. Furthermore, unions had begun to influence hiring and firing. Thus the initial success of the workers' revolution set off a process through which mill hands learned to adapt their workplace strategies, at each step demanding more control for unions and less for owners and administrators. The outcome of this militance was that the victorious Constitutionalists had no choice but to write the most proworker labor law in the Americas. That law is the subject of the next chapter.

In this regard, a conflict in mid 1916 in the large Río Blanco mill, shortly before the opening of the Constitutional Convention, illustrates the dilemma confronting owners and new elites. On July 11, the union held a meeting to deal with "Asunto Toquero." Two unionists accused

Toquero of snitching on them to foremen. Toquero defended himself, saying that he *"nunca ha sido contrario a sus compañeros los obreros."* The union leader then judged Toquero innocent and "authorized him to continue in his employment.[91] If he had been judged guilty, Toquero would have lost his job. By 1916, the larger and stronger unions had acquired the unprecedented power to determine who worked, or did not work in the mills. The law would not be able to ignore such power, the result of "efforts carried out with weapons."[92]

CHAPTER 6

The Institutionalization of the
Labor Regime: Law and Government

GIVEN THE VIOLENCE IN THE MILLS and the surprising new power of workers, institutionalization of labor affairs became inevitable. This institutionalization had two main thrusts, labor law and trade unions. Chapters 4 and 5 described the beginnings of the former in the 1912 agreements and the military decrees. This chapter takes that process through the 1917 Constitution and subsequent state labor codes. Chapter 7 will look at the formalization of trade unions.

As early as Madero's government, there was an increasing consciousness of the need to quell the labor revolt through labor law:

Because of the country's recent political change, there have developed a number of social movements. Strikes have followed upon strikes, and the government, worried about this state of affairs and wishing to find a solution, agreed to convoke this meeting with the object of discussing the bases to regulate labor in the Republic, which the government proposes to elevate to the category of law, presenting initiatives to the Federal Congress.[1]

In that early period and subsequently, labor law included the development of three interrelated sets of formal institutions—law, collective contracts, and government labor offices—none of which existed in 1910. Except for a few large factories with brief written work rules, most workplaces relied on personal relationships, customary behaviors, and the authority of ownership. The expectation was that the boss ordered and the worker obeyed. Factories hired, fired, and disciplined workers at will. When Gómez Abascal posted the 1904 work rules in El Carmen, one of them was "any complaint will be useless."[2] There were

no collective contracts, no body of labor law, and no government offices specifically dedicated to labor issues.[3] In short, the labor regime
was mostly informal, and what formal components existed—the Constitution, commercial law, and the laws of property—favored owners
over workers.

The workers' revolution completely changed this. Between 1910 and
1923, Mexico created the most elaborate and progressive labor relations system in Latin America. By the end of the revolution, there existed a very complex labor relations system with negotiated work rules;
collective contracts; detailed federal and state labor laws; and numerous federal, state, and municipal government labor offices. Previous
chapters explained the origins of the new system. This chapter concen-
trates on its maturation through the revolution and into the early institutional period.

While legal history can often appear dreary to the social historian,
particularly those interested in revolution, the importance of this chapter cannot be underscored enough. How Mexico created, from nothing, this set of proworker labor institutions in just a decade has been a
mystery, except for the false claims after the fact that it came from
above, from the state. Furthermore, the legal framework established
during the revolution—Article 123, detailed and progressive labor
codes, powerful government labor offices tied to the unions—became
a singular factor of Mexican politics, economic development, and social process through the rest of the twentieth century. In fact, it would
be as difficult to explain Mexico's twentieth-century history without
this as it would be to ignore the agrarian revolution and later expropri-
ation of the great estate.

Equally important are three final concerns. First, the progressive nature of Mexican labor law in light of a labor market whose relations of
supply and demand continued so unfavorable to labor suggests the
depth of the workers victory in their revolution. Law was labor's victory and must be understood in this light. Second, a corpus of law that
both favored and controlled trade unions illustrates the advantages
and disadvantages of a revolution led by such organizations. Unions
ultimately could not exist without the new law, which is to say without the state, which is why the workers' revolution had the ambiguous
outcome that it had. Third, every successful revolution must create a
new hegemony or die. While part of that hegemony is undoubtedly
cultural, another part is institutional and legal: labor law.

TABLE 6.1

Important Sections of Article 123

I and II—Established the eight-hour day, seven for the night shift.

III—Regulated hours of work for minors.

IV—Made Sunday an obligatory day of rest.

V—Extended paid leave for mothers with newborns to one month, with two half-hour periods when they return to work to breastfeed their babies.

VI—Established a legal minimum wage. "The minimum wage to be received by a workman shall be that considered sufficient, according to the conditions prevailing in the respective region of the country, to satisfy the normal needs of the life of the workman, his education and his lawful pleasures, considering him as the head of a family."

VII through XI—Mandated other aspects of wages. Section VII required equal pay for equal work, specifically requiring employers to pay women the same as men. Section IX instituted minimum wage commissions in each municipio, subject to state Labor Boards.

XII through XV—Mandated housing and health benefits. Section XII required all businesses employing more than 100 workers to provide comfortable and clean housing for their workers. Section XIV made employers responsible for work-related accidents and illnesses.

XVI through XXII—Permitted workers to unionize and allowed owners' associations. Section XVIII defined legal strikes as those whose goal was to "achieve the equilibrium between the diverse factors of production." However, strikes were only illegal when the majority of strikers engaged in violence or when workers in military establishments struck during wartime. On the other hand, Section XIX made owners' lockouts illegal, except when an excess of production made it necessary to shut down, and then only with previous approval of the Junta de Conciliación y Arbitraje.

XX—Ordained that "Differences or disputes between capital and labor shall be submitted for settlement to a board of conciliation and arbitration to consist of an equal number of representatives of the workmen and of the employers and of one representative of the Government." Section XXII provided that workers fired for joining a union receive three month's salary as compensation. Sections XXIII–XXV dealt with the mutual debts of workers and owners, providing legal protection to workers. XXV provided for free employment services.

XXVI—Protected Mexican workers who signed labor contracts with foreign owners.

XXVII—Contained the eight clauses that would make a labor contract null and void, protecting the rights of workers.

XXIX and XXX—Provisions for housing, retirement, unemployment, and disability funds.

SOURCE: Constitución Política de los Estados Unidos Mexicanos, firmada el 31 de enero de 1917 y promulgada el 5 de febrero del mismo año (Mexico, 1917), 38 pp.

Legal Change, 1911–1912

The intense beginning of the workers' revolution forced Mexican elites to consider new strategies to contain the suddenly insurgent proletariat. Confronted with the failure of repression, they began to create a formal labor relations system. On September 22, 1911, interim president Francisco León de la Barra took the first step when he proposed the establishment of a Labor Office that would "Arrange fair regulation in cases of conflict between owners and workers, and serve as arbiter in their differences, whenever they might solicit its services."[4] The Congress approved the Labor Office, and it started up on December 18, 1911. The date was significant because it was after Madero became president but just three days before textile workers launched their great strike.[5]

The Labor Office immediately established itself as the center of the formal labor relations system and grew steadily through the twentieth century. Originally part of the Secretaría de Fomento, Colonización e Industria, in 1917 the Ley de Secretarías de Estado made it a department within the new Secretaría de Industria, Comercio y Trabajo. It became an autonomous department in 1932, then the modern Secretaría del Trabajo y Previsión Social in 1941.[6] Under President Madero, it provided government its first institutional mechanism to intervene in what had previously been considered a private matter, the relations between owners and workers, and it continued in that role through the rest of the century.

Its first director, Antonio Ramos Pedrueza, presided over a small staff that collected data on working conditions, workers, unions, and conflicts. The Labor Office's inspectors traveled to factories, interviewed owners and workers, and sometimes provided political and military intelligence as well as advice on labor affairs.[7] Ramos Pedrueza himself played a central role in the 1912 textile convention, receiving criticism for acting more like a cabinet member than a subaltern.[8] When Huerta replaced Madero, Adalberto A. Esteva replaced Ramos Pedrueza. Surprisingly, Esteva's outlook paralleled that of Ramos Pedrueza, noting that "Solving the worker problem, as well as the agrarian problem, will provide stability to the Mexican government, making impossible the revolutions."[9] Subsequent twentieth-century Mexican presidents followed Esteva's advice.

The second step in the institutionalization of the labor relations system was the de facto tripartite—owners, workers, government—

convention of July 1912. The new Labor Office officially convened a meeting of owners, but unofficially invited labor leaders and gave them authority to effectively veto the final agreements. Thus was born the tripartite convention, a staple of twentieth-century Mexican labor affairs. The Labor Office believed the convention a great success because it provided "a notable benefit for the resolution of the differences that arise between owners and workers and the greater peace between them."[10] A second tripartite convention of the textile industry met from October 1925 until March 1927, drafting a new contract law to replace 1912 agreements. Subsequent Mexican governments continued using tripartite conventions as the preferred method to draft contract law in textiles and other major industries.[11]

The 1912 agreements, the Reglamento and Tarifa, represented the third step. The convention did not label them a "collective contract" because Mexican law at that time lacked any reference to nonindividual work agreements. Nonetheless, they contained the basic elements of collective contracts: wages and working conditions. However, without further legal change, they remained voluntary agreements enforced by a partial exemption from a government tax. Marjorie Ruth Clark was correct in noting that the agreements were "the first in Mexico which had even a semblance of collective bargaining."[12]

Despite this institutional progress, industry lawyer Tomás Reyes Retana commented that "we are still children in these industrial matters."[13]

The 1917 Constitution

After 1912, the political situation in Mexico deteriorated so greatly that it became impossible for the federal government to pursue further institutionalization. In fact, the collapse of the Huerta government was the collapse of central government itself. The warring factions that survived nonetheless had to deal with conflict in whatever areas they controlled. When Constitutionalist military commanders took power in Puebla and Veracruz, they discovered that they needed to constrain conflict in the mills if they were to exercise power. This forced them, as it had Madero and Huerta, to continue the institutionalization of labor affairs, albeit regionally while ceding to workers much of what they had already obtained through violence.

The military decrees attempted to control the workers revolution through a combination of giving to workers what they demanded—a minimum wage, higher overall wages, lower hours of work, benefits, recognition of unions—while also providing government with the tools to regulate labor affairs, particularly labor offices to monitor unions and strikes. The Puebla decrees established a state labor office, a state labor court, and state labor inspectors. In Veracruz, the decrees provided for state labor inspectors and gave municipal governments authority in labor affairs. These offices attempted to regulate conflict but also provided legal recognition and protection to trade unions. Since the prevailing Constitution, written in 1857, considered trade unions an illegal restraint on trade, the accumulation of institutional changes between 1911 and 1916 signaled what the Constitutionalists would do if they obtained power: write labor law that both strengthened workers and the ability of the state to control them.

By late 1916 the Constitutionalists controlled enough of the country to impose their legal and political ideas. On September 14, Carranza decreed a constitutional convention. He quickly organized elections for convention delegates, who began work on December 1.[14] They finished on January 31, 1917, and the new Constitution took effect on February 5.[15] The convention used the 1857 liberal Constitution as a model but added two completely new chapters, Article 27 on social property and Article 123 on labor. Clark wrote, "many of the provisions pertaining to labor only pretended to legalize and generalize conquests already made by different groups in the working classes. Both articles 27 and 123, although startling to the outside world, had been foretold by decrees issued by Carranza or by other 'generals' in the pre-constitutional period."[16] Two of these generals and representatives to the Congreso Constituyente had already written military labor decrees, Heriberto Jara and Cándido Aguilar.

There was no labor clause in the 1857 Constitution. Its authors wrote a document that responded to the conflicts of the early republic when Mexico suffered numerous foreign invasions, the loss of half the national territory, and the inability of the country's elites to reach a consensus on a social, economic, or political project. The political chaos of the early years of the Republic attested to the difficulties of nation building. When the liberals took power after the Plan de Ayutla, they needed to save a nation rather than protect the underclasses. Their Constitution therefore addressed the state rather than the people. It unleashed a decade-long civil war among elites for control of that state that culminated in the absolute victory of the liberals.

The liberals had little to say about workers other than the core liberal idea of creating and protecting a free market in labor. On labor, they labeled Title I, Section I, "Of the Rights of Man," in clear allusion to the revolutionary French document of 1789.[17] In France, "the Revolutionaries believed in the free market, that all must be free to trade on equal terms."[18] That was the extent of legal ideas on work in mid-nineteenth-century Mexico. Article 2 of the 1857 Constitution ratified the complete abolition of slavery. Article 4 declared that "Every one is free to engage in any honorable and useful profession, industrial pursuit, or occupation suitable to him, and to avail himself of its products." Article 5 affirmed that "No one shall be compelled to render personal services without due compensation and without his full consent."[19] Nineteenth-century Mexican liberals understood that the nation-state is a set of institutional arrangements that define the free market and its rules. To construct a modern nation state, they needed to define and protect the market. Consistent with this view, the Penal Code of 1872 "punished with arrest and fine 'those who make a tumult or riot, or employ any other methods of physical or moral violence for the purpose of raising or lowering the wages of the workers or who impede the free exercise of industry and labor.' "[20]

Between 1910 and 1917, Mexico's liberal project collapsed. Although Madero, Carranza, and other wealthy landowners had little disagreement with the free market, the revolution from below was a vast upheaval against that market and its ideology. From Juárez to Díaz, the liberals brought to Mexico an increasing concentration of land and wealth. The results devasted rural communities and poor people. That is why, from Morelos to Chihuahua, poor rural people poured into the revolutionary armies. This was the case with the initial Maderista revolution and also with the later Zapatista, Villista, and other revolutionary armies. For many poor Mexicans, the free market had brought ruin and they wanted something else.

While wealthy Mexicans profited more from free-market economic growth, many feared that such policies favored foreigners over Mexicans. Growing nationalism tempered their belief in the market. Industrial workers experienced free markets as the right of factory owners to control their jobs and their livelihoods, which for many meant degraded working and living conditions and an affront to their dignity. It also meant a labor market in which supply and demand favored owners over workers. For these reasons, by late 1916 the ideology of laissez faire capitalism had collapsed. Furthermore, by that time the underclasses were armed, dangerous, and hostile to the forces that had overwhelmed their lives. In cotton textiles, "During the year 1916 and part of the next, there was a deep agitation, and strikes followed strikes and the textile industry only worked with great disorders."[21]

The delegates to the Constitutional Convention responded to these challenges though two unique and antiliberal additions to the 1857 Constitution. Article 27 mandated that "original ownership of land and water in Mexico belongs to the state." The state could cede property rights to private citizens although it "shall at all times have the right to impose on private property such limitations as the public interest may demand."[22] Although the article drew upon Spanish legal traditions rather than socialist ideology, it addressed the property concerns of most revolutionaries. It undermined the liberals' idea of the inviolability of private property, allowing the state to override the interests of wealthy Mexicans and foreigners.[23] However, whether the state would actually take that step would be determined by politics and not law.

Article 123, "Del Trabajo y de la Previsión Social," was the second of the antiliberal provisions. It mandated a set of labor rights that were radically new in Mexican, indeed, in Latin American constitutional law.[24] The lengthy article built on the legal work of the generals and colonels who between 1914 and 1916 legalized unions, mandated minimum wage and work provisions, established workers benefits, and provided for government intervention in labor affairs. As Gruening noted, "The labor provisions of the Constitution of 1917 perfectly reflect past grievances."[25] Its thirty separate items covered hours and wages, fringe benefits, unions and strikes, and other issues important to Mexican workers.

Many observers have commented that law and constitutions have a different meaning in Mexico than in the United States and not only because of the Hispanic and Napoleonic origins of the Mexican legal system. In the United States, the Constitution establishes fundamental law; in Mexico, it sets fundamental goals. Law in the United States is mostly observed while law in Mexico is often ignored. Despite the validity of these objections, it would be a mistake to ignore Article 123.

First, whatever the defects of law and legislation in postrevolutionary Mexico, the Constitution sets standards that any citizen can claim as just. Article 123 proclaimed to all Mexicans that workers now had rights previously ignored: to organize unions, to strike for better working conditions, to demand a decent standard of living. Before 1917 these rights could not be violated because they did not exist. Now, even if everybody violated them, they were still fundamental rights of workers according to the ultimate law of the land. This shift from a complete absence of constitutional rights for workers to many and progressive rights would have been inconceivable if the Congreso Constituyente had not observed persistent conflict in the mills. The delegates wrote Article 123 to be a solution to a problem, which was combative workers. In this regard, Article 123 was a victory for the factory revolt.

Second, Article 123 legitimated workplace standards for Mexican workers. That the standards were not absolute but subject to constant negotiations as inflation, technological change, and market shifts affected the economy of industry did not diminish their importance. It simply meant that the most important Constitutional provisions were those concerning the conditions of negotiation rather than specific standards themselves. In this area, Article 123 ratified what cotton textile workers had won in practice and then in state military decrees: strong unions and effective strikes. Section XX formalized what state military commanders had discovered during the civil war: final authority in the factory no longer belonged to the owners. "Differences or disputes between capital and labor shall be submitted for settlement to a board of conciliation and arbitration to consist of an equal number of representatives of the workmen and of the employers and of one representative of the Government." If this left workers not controlling the workplace, it nonetheless weakened the authority of owners. The 1917 Constitution ratified the power that workers had conquered in practice. Once again, law followed practice.

Third, Article 123 established the framework for a new set of labor organizations in Mexico. While law could be ignored, the organizations could not. The Constitution provided for worker and employer associations; government inspection and arbitration; and federal, state, and municipal labor bureaucracies. These organizations existed, acted, and constituted a new industrial relations system in the country. For example, in 1918, Puebla Governor Alfonso Cabrera decreed state municipal wage commissions with equal representation of owners and workers to carry out the Article's mandates on minimum wages and profit sharing. To date, a national minimum wage commission fixes the legal minimum wage in Mexico.

Although Article 123 established progressive labor legislation, it was a compromise. Conservative liberals only grudgingly conceded the need to protect labor at all, while radical liberals and Jacobeans were convinced that there could be no social peace in Mexico without legal justice for the working classes. It was an unsatisfactory compromise to many, reflected in the Article's notorious lacunae. Mexican Constitutional law requires complementary statutory and regulatory law without which Constitutional precepts are a dead letter. Although Carranza wanted to federalize labor legislation, the Congreso Constituyente left to the states the execution of Constitutional mandates through state labor codes. The federal Constitution therefore established certain minimum goals and rights for workers but left no practical means for them to be implemented.

State Labor Codes: Veracruz and Puebla

Nonetheless, Article 123 did not become a dead letter. Instead, the states implemented it in such a radical fashion that it eventually (1931) became necessary for the federal government to impose a national labor code. As we will observe in Chapter 8, the state codes drafted in the decade following the Constitutional Convention became radical for the same reasons that Article 123 was radical: a continued fight for authority at work by combative and strengthened workers. Mexican state governments issued state labor codes to implement Article 123, beginning with Yucatán and neighboring Tabasco in July and September 1917.[26] Other states implemented labor codes between 1918 and 1926: Campeche (1918 and 1924), Nayarit (1918 and 1924), Sonora (1919), Sinaloa (1920 and 1926), Puebla (1921), Chihuahua (1922), Querétaro (1922), Jalisco (1923), Michoacán (1923), San Luis Potosí (1923), Nuevo León (1925), Tamaulipas (1925), and Veracruz (1918, 1921, 1923).[27] The most important and influential of the early codes was that issued in January 1918 by Veracruz governor Cándido Aguilar, author of two military labor decrees and representative to the Congreso Constituyente. A large state with a highly conflictive textile industry and powerful textile unions, Veracruz's code became a model for other states.[28] With eight titles or divisions and 206 articles, it was considerably longer and more detailed than Article 123, from which it received its mandate. It broke new ground for Mexican labor law by creating unprecedented rights and benefits for workers while also providing government new controls over unions and the labor relations system.

"Title One" defined the labor contract. Article 21 recognized collective contracts and the right of unions to be signatories to them. Although the military decrees and Article 123 of the federal constitution hint at collective contracts, Article 21 of the Veracruz labor code specifically provides the right to collective contracts and the right of an agrupacíon de trabajadores to be party to them. Article 22 limited that right to unions "legally constituted according to this law." Since prerevolutionary Mexico was innocent of both collective contracts and legally recognized unions, these articles created in law what textile workers had achieved in practice: a new labor relations system based on unions, collective bargaining, and collective contracts.

Article 33 established eleven "common obligations of all owners or in their absence, their administrators or representatives to the workers."[29] These included preferring Mexicans over foreigners; treating workers "with due consideration, abstaining from treating them badly by word or by deed"; providing one-month salary each year as profit sharing; and "Listening to workers' complaints about the employees [supervisors] and correcting the problems caused." Two of these items deserve special note. The extra month salary in Section Seven became the profit sharing mandated by Article 123. As other states and then virtually all employers adopted this, it became the famous and virtually universal aguinaldo that Mexican workers enjoy to the present. Second, the law mandated that companies correct the abuses of supervisors, which strengthened workers at the expense of management.

Article 34 prohibited bosses ("todo patrón, jefe, empleado o maestro de fábricas, talleres y demás establecimientos") from mistreating workers:[30] It specifically prohibited employers from retaining wages for fines, forcing workers to buy items in specific stores (the tiendas de raya), demanding or accepting money from workers as payment for access to jobs, charging interest for wages paid in advance, and forcing workers "through coercion or any other means" to leave the union. It also prohibited employers and their representatives from showing up drunk, carrying arms, or collecting money from workers inside the factories. Also, it prohibited "Any other action or abuse that might result in prejudicing workers or the freedom of action."

Article 35 covered nine parallel worker obligations, including "behaving well and treating the owners and his representatives with consideration and due respect." Although it required workers to obey factory work rules, it stipulated that the Junta Central de Conciliación y Arbitraje del Estado had to approve the rules.[31]

These articles were not just law, they were the legal overthrow of the old labor regime in Mexican factories and the installation of a new one. Before Mexico's revolution, owners and their supervisors ran the factories as they wished, subject only to common cultural expectations.

Owners wrote the set of written obligations that workers had to carry out in the factory. The 1912 tripartite convention of the industry introduced some minor limits to owner work rules. However, the deepening violence inside and outside the factories after 1912 brought more worker influence than anybody in Mexico had ever thought possible. The military decrees attempted to tame the violence but the continued challenge to authority in the factory pushed the revolution forward, reflected in Article 123 of the 1917 Constitution. By 1918, the Veracruz labor code legalized, formalized, and institutionalized what had happened in practice: the owners were no longer able to command at will. The Veracruz code specifically and strongly endorsed unions and collective contracts, absolutely prohibited companies from firing workers for belonging to unions, forced factory owners to listen to workers' complaints and to correct wrongs, and unconditionally prohibited companies from physically, verbally, or financially mistreating workers. Furthermore, although Article 35 obligated workers to obey factory work rules, it made the factories register them with the state government, tantamount to allowing an outside authority to intervene in the process. This was not trivial. The set of rules contained in Articles 33, 34, and 35 took away from the owners their absolute dominion over the factory that had prevailed during the Porfiriato. Although it did not directly transfer authority to workers, it provided fundamentally new protections for workers while placing the state in the middle of the factory through its power to approve work rules. The legally increased power of workers and the state in the factories came at the expense of management.

Article 40, section 4, went further. It required owners of large industries (more than one hundred workers) to have written work rules. Article 41 decreed that the reglamento be drafted by two representatives of the owners and two of the workers, then submitted to the Junta Central de Conciliación y Arbitraje del Estado for approval.[32] Thus, medium and large factories could no longer write their own rules. The law could not be more clear: everybody—owners, workers, and the state—participated in running the factory. This was not socialism, but it was a new limit to the definition of private property rights.

Chapter Nine of Title One, "Of the Termination of the Contract," set the rules for ending the labor contract, effectively limiting the right of owners to fire workers. Most important was Article 96, which only permitted firings for specific reasons:

1. Section 1 permitted firing workers when they did not do the agreed-upon work, "in the judgment of the Boards of Conciliation and Arbitration."

2. Section 2 allowed termination when the workers did not obey orders, "according to the judgment of the same Boards."

3. Section 5 determined that a worker could be discharged for violating three times in a month the "reglamento interior de la fábrica."[33]

4. Section 6 allowed owners to discharge workers who damaged their interests "a juicio de las Juntas de Conciliación y Arbitraje."

Meanwhile, Article 97 prohibited termination for joining a union or participating in a strike.[34] These provisions forced owners to appeal to state labor boards in order to fire workers even if the latter did not do their work, refused to obey orders, or assaulted supervisors. It specifically permitted workers to violate factory rules twice a month without fear of firing! It protected the unions that assuredly would defend the workers before the state labor boards. The severe constraints on firing, the main disciplinary tool for employers, shifted authority in the workplace away from owners and toward unions and state labor boards.

Title Three implemented the wage aspects of Article 123. Chapter Two defined the minimum wage as that necessary to "satisfy the normal necessities of the worker's life, individually or considered as the head of a family."[35] Following guidelines laid down by the Constitution, it provided for the establishment of Municipal Comisiones Especiales del Salario Mínimo. With instructions from the Junta Central de Conciliación y Arbitraje del Estado, local commissions would determine the minimum wage in each municipio. Workers (in reality, unions) and owners in each industry sent an equal number of representatives to the Comisiones Especiales, guaranteeing legal parity.[36]

Title Five, "De los Sindicatos y Federaciones," defined the legally constituted unions that could be party to a collective contract. The law required that unions register with the municipal government, forwarding the minutes of their initial meeting, the members of the leadership, a copy of the bylaws, and monthly reports. Businesses could not refuse unions, but unions could not coerce workers into joining. The law disallowed "in their membership agitators or people that carry out propaganda of subversive ideas."[37]

With this clause, the law made unions legally powerful but potentially subordinate. The Veracruz labor code recognized and defended unions but subjected them to requirements that could subordinate them to strong governments. Only legal unions could sign collective contracts and only the state could make a union legal. Furthermore, the state prohibited unions from affiliating workers who held antistate ideas, the practical meaning of ideas *disolventes*. The specific institutions that the law created to control unions and business were the Juntas Municipales de Conciliación and the Junta Central de Conciliación y Arbitraje del Estado. The Municipal Boards were composed of two representatives of the owners, two of the workers, and the Sindico of the Municipio. The State Board consisted of three representatives of the owners, three of the workers, and one representative of the state government.[38] Assuming that owners and workers would normally oppose each other, effectively the law allowed the state to decide.

The Juntas had the right to approve factory work rules and decide on conflicts between owners and workers. In case of conflict, the municipal boards served the role of conciliation in the first instance. If that failed, either party could take the issue to the state board, which offered arbitration and conciliation. Arbitration was not subject to the formal rules of Mexican law.[39]

The Veracruz code included various provisions to protect women. These clauses prohibited women from taking jobs that required them to work late at night as well as dangerous or unhealthy jobs. It also reinforced equal pay for equal work (Article 93) while providing eight weeks of maternity leave at half pay (Article 91).

In Veracruz, the combination of federal and state labor law created numerous organizations that intervened in labor affairs. A complex hierarchy of municipal, state, and federal labor boards provided workers numerous opportunities to file grievances or challenge owners. What owner would fire an incompetent laborer if a union defended him or her in this process? The cost of the various appeals, the time delays, the procedural difficulties, the notorious slowness and corruption of government offices, the equality of vote between unions and owners, and the uncertainty of the final decision robbed the owners of their prerevolutionary capacity to run the factories as they wished. Instead, workers and unions had acquired new power in the workplace, as did a state fortified by numerous labor institutions. Owners no longer controlled workers.

In October 1917, the governor of Puebla, Alfonso Cabrera, issued a law to bring the state into compliance with Article 123. The decree's three articles did little other than establish a state labor board with broad powers, "whose object is to resolve all conflicts between Capital and Labor in the State."[40] Four years later on November 14, 1921, General and Governor Jose María Sánchez signed the state's new labor law, a massive document with 110 pages and 330 articles.[41] In basic structure and outline, it showed the influence of Veracruz.[42] Each of the ten titles regulated a major aspect of the labor relations system:

Title One—The labor contract
Title Two—Maximum workday and minimum wage
Title Three—Factory and business work rules
Title Four—Unions
Title Five—Strikes and lockouts
Title Six—Profit sharing
Title Seven—Accidents and illnesses
Title Eight—State labor department
Title Nine—Health and safety
Title Ten—Miscellanea

Thus in the four years since the federal constitution, state-mandated benefits, state-protected unions, state-enforced collective contracts, and government regulation were becoming standard legal fare. Veracruz had defined most of these items, and Puebla broke little new ground except in one significant area.

Like Veracruz, Puebla law defined and recognized collective contracts, unions, and the common obligations of workers and owners. It added the requirement that state government approve factory work rules. It also mandated that a commission in which unions and owners had the same number of representatives draft the rules (Article 158). The state code limited the right to fire to four very specific reasons, each of which could be appealed to a set of labor boards: Junta de Conciliación Permanente, Juntas de Conciliación Temporal, Arbitros, and the Junta Central de Conciliación y Arbitraje. Once again, a state government made it legally very difficult to fire a union worker.

The significant change with respect to Veracruz was in hiring, which in the former state was left to owners. In Puebla, Articles 107, 109, 110, and 112 mandated and protected unions and gave them, rather than owners, the power to hire workers. Workers who wanted a union had the right to a collective contract, and the union, not the company, "queda facultado para sustituir a los trabajadores en la prestación del servicio" (Article 107). Articles 109 and 110 further strengthened the role of the union in hiring workers. Article 112 required that a business or factory sign a collective contract with its workers if a legal union existed.[43]

Labor Law and the Labor Regime

Law is a set of formal rules that may or may not be followed. Legal reality is one world and what actually transpires on the shop floor is often another. Even with these caveats, legal change in Mexico between the outbreak of the workers' revolution in the general strike of 1911 and the Puebla state code of 1921 is stunning. In 1910 there was no labor law. By 1921 a combination of federal and state law provided textile workers:

1. The right to unions.
2. The right to strike.
3. Strong protection against firings.
4. Equal representation of unions and owners on the labor boards that heard appeals.
5. Equal representation of unions and owners on commissions that wrote factory work rules.
6. The requirement for state governments to approve factory work rules.
7. The right to minimum and overtime wages and maximum hours of work.
8. The right to Sundays and numerous holidays.
9. A vast array of education, housing, maternal, and medical benefits.
10. Union hiring (Puebla).

These were not inconsequential rights, and their real support came from workers who had fought for them throughout their revolution. Thus, for example, in the middle of a minor wage dispute in Santa Rosa in late 1917, against company wishes, the union demanded that women workers receive equal pay for equal work, citing the new Constitution. Women had fought alongside men during the revolution, had achieved some equality before the law, and by 1917 male-dominated unions demanded that companies comply.[44]

On firings, the rules determined that when a mill decided to fire a worker, the dismissal remained subject to appeal through various labor boards on which unions had the same votes as the owners. There were municipal, state, and federal boards. Workers remained on the job with pay until the appeals process was complete. Normally the worker had to violate a procedure three times to even initiate a legal dismissal.

Imagine the process. During the revolution, workers often struck supervisors, and even more often refused their orders. With these laws in place, only after a worker hit his boss a third time could the mill initiate a dismissal. The union would naturally appeal to the municipal labor board. If in Atlixco or Orizaba, the municipal president was probably a union man and it is likely that the board would rule in favor of the union, so the aggressive mill hand retained his job. Meanwhile, the union hired new workers, formally in Puebla and informally in Veracruz and elsewhere, which meant it hired loyal unionists. Disciplining a worker was subject to the same process as dismissal, so that in law the factory had no mechanism with which to control the shop floor.

These were unprecedented rights for workers, and they were mandated and regulated by state law under the authority of the federal Constitution. If the workers' revolt prior to 1921 sought to improve working conditions and to strengthen the role of workers in running the workplace, by 1921 the law gave them their victory. Although they had achieved many of these gains in practice prior to the new laws, the federal Constitution and state labor codes gave them a permanency they otherwise would not have had. For workers and for unions, this was not inconsequential. The new law meant that the new state recognized their victory. Yet victory in law was only half of the institutionalization of the proworker labor relations system. The other half were powerful trade unions mandated and protected by these laws.

CHAPTER 7

The Institutionalization of
the Labor Regime: Unions

IF THE RADICAL VICTORY of the workers' revolution can be seen in labor law going from nothing in 1910 to an extensive set of proworker labor codes in 1923, an equally fundamental shift can be seen with trade unions. Outside the law in 1910, by 1923 unions controlled all jobs in the textile industry. They signed collective contracts, and their leaders were among the most powerful men in Mexico. The most powerful union boss of all, Luis N. Morones, was one of the country's four or five most influential political figures in the 1920s. Indeed, it would be difficult to overstate the importance of organized labor after the revolution because the postrevolutionary state relied on the CROM, first, and then its successor, the Confederación de Trabajadores de Mexico (CTM), afterwards. The power and pervasiveness of trade unions extended through business and industry, daily and family life, popular culture, and even the high art of the Muralists. In textiles as well as other industries, local labor leaders replaced the old Porfirian political bosses, and powerful unionists like Martín Torres in Orizaba and Antonio J. Hernández in Atlixco ruled their towns with an iron fist. Industrialists who once commanded now yielded to their former enemies, the unions.

The change in the role of unions is not surprising because the workers' revolution was, above all, a trade union revolution. Mill hands chose unions as the primary vehicle for their revolt in the factory. They took an organizational form with which they had acquired long familiarity during the previous decades of industrialization and used it to press their workplace demands: wages and hours of work, control over hiring and firing, control over the rhythm of work on the shop floor, benefits tied to employment in the industry, control over the community, and even control over foremen.

Along with a change in power came an inevitable and fundamental transformation of the trade union form, particularly in the internal relationships between union leaders and rank-and-file workers. This chapter traces the changes in trade unions from 1910 to 1920. The sometimes ignored and usually combated unions of the late Porfiriato were a far cry from the wealthy and powerful organizations that came to control the shop floor and the mill towns after the revolution. The shifts can be best understood by looking at textile unions as they advanced through four stages: (1) Porfirian unions; (2) from the opening of revolution to the Textile Convention, 1910–12; (3) revolutionary consolidation, 1913–16; and (4) the origins of organized labor, 1917–20. In each phase, trade unions were both protagonists and products of the social events that created modern Mexico.

Porfirian Unions

The modern word for trade unions in Mexico is *sindicato*. With well-defined legal, cultural, and social meanings, the word is clearly understood in modern Mexico. Before the revolution, however, workers rarely used *sindicato* to describe their organizations. More commonly, they called them *asociaciones* (associations), *gremios* (guilds), *sociedades* (societies), or *cooperativas* (cooperatives). The reasons were both historical and legal. Historically, workers and small producers had acquired familiarity with guild and mutual associations during Mexico's pre-industrial history, so it was natural to view more modern organizations as continuations of the older groups. Legally, Mexican law did not recognize unions, so there were no legal names or common public labels. The varied terms matched the varied aspirations of workers.

Owners, however, were not all confused about labor organizations. They opposed them. Although they sometimes tolerated worker self-improvement groups, they would not allow any interference with the owners' rights to set wages, hours of work, discipline, and work standards. Since Mexican law did not provide for collective contracts, owners and government could easily repress workers' organizations. As a consequence, most workers' associations walked a fine line between defending workers and not antagonizing owners.

After the revolution it became common to write about Porfirian repression of unions, but the actual history of labor organization was a bit more complex. Some workers organized with varying degrees of success, carrying out strikes and developing a sense of solidarity. Without this previous history of unionization, the later workers' revolution would never have been possible. Nonetheless, no union could be party to a collective contract or enforce discipline within the organization or the factory. The prevailing legal environment made participation in labor organizations voluntary and risky. In the textile industry, some mills organized and some did not, but most labor organization was short lived.

Despite legal and political problems, in the 1860s and 1870s mill hands experimented with mutualist societies. In 1872, a group of militants founded the Círculo de Obreros de México.[1] A successful strike in La Colmena led to the Asamblea General de Obreros Textiles del Valle de México in 1873, which gave rise to the Unión de Resistencia de Tejedores del Valle de México.[2] By the end of the century, most large mills had a mutualist society, which in times of conflict or strikes could function as a union. Textile workers carried out at least 30 strikes between 1881 and 1895, with strike waves in 1881, 1884, 1889, 1890, 1891, and 1895.[3] In 1884, the Puebla textile organizations formed a regional labor federation, the Federación Obrera.

During the latter half of the Porfiriato, the strikes became larger. When they became dangerous, authorities intervened to mediate or to repress. For example, in 1898 mill hands in La Colmena and El Barron (Mexico State) struck when the owners lowered wages, as did those in El Mayorazgo (Puebla) in 1900 and La Hormiga (Mexico City) in 1901. In the State of Mexico strike, the governor mediated but the workers lost. The strike in El Mayorazgo was joined by other Puebla City mills but the owners won because large supplies in the warehouses outlasted the hungry families. Only in La Hormiga did the workers win, because the local political authority sided with them.[4]

The consequence of this activity was that by the late Porfiriato, many textile workers had experience with organization, militancy, and strikes. Although they won some of their conflicts, they usually lost because the law did not recognize collective labor agreements. Under such conditions, no trade union could achieve lasting success.

In 1906 textile workers founded the Grand Circle of Free Workers. Influenced by anarchists and the Flores Magón brothers, the organization declared "War to the death to the tyrant who sells us, to the merchant who robs us, and to the employer who exploits us; war without quarter . . . by reason or by force."[5] With branches in the traditional textile centers of Orizaba, Puebla, and Atlixco, it played an important role in labor conflict that year. In August, Pascual Mendoza founded the Gran Liga Nacional Obrera "Esteban Antuñano" in Puebla.[6] In October, Mendoza and José Morales founded the Segundo Gran Círculo de Obreros Libres (GCOL) in Atlixco, with branches in El León and Metepec.

The new organizations made radical demands, to which the Puebla textile owners responded by founding the CIM. The CIM quickly issued work regulations "intended to halt increasing labor agitation."[7] These regulations led to the famous labor conflict of 1906 and 1907. With the slaughter in January 1907, the owners imposed their will, at least for the time being. In Atlixco, the leaders of the new unions fled the town, and the GCOL branch there met its demise.[8] Some worker organization subsisted, however, and owner Rivero Quijano noted that, "Naively, some believe that we have definitively finished off labor organization."[9]

In Atlixco and Orizaba, labor activists moved quickly to reorganize after the defeats of 1906 and 1907. Although we lack a full history of the process, we know that workers in the larger mills formed asociaciones and sociedades, carefully working with local authorities to avoid a confrontation that would lead to dismantling their organizations. In late 1907, Metepec mill hands organized the Círculo Fraternal de Obreros, showing their caution in a letter to the local Jefe Político: [10]

Citizen Jefe Político,

We who write, workers of the factory established in Metepec where we are neighbors and up to date with our taxes, make manifest: that, making use of the right of association in Article Nine of our very free Constitution . . . we have formed a Society whose name is "Círculo Fraternal de Obreros . . ." As you can see, Señor Jefe Político, everything is within law and order . . . we will never earn the suspicion of the authorities that rule us.

In December 1908, Pánfilo Méndez informed the Jefe Político of Río Blanco that they had established a "Mutualist Savings Society" in the local mill.[11] He wrote that their goal was "mutual aid with life's difficulties, individual and collective savings, and mutual protection in the most urgent cases. As you can see Señor Jefe Político, everything is within law and order and we swear our respect and security."[12] Despite such letters, local political bosses were always suspicious of labor organizations, used secret informants to spy on their proceedings, and often brought leaders before local judges if there were rumors of a strike or other sort of labor action.[13]

Although some labor historians have not classified these organizations as trade unions, they in fact carried out many of the functions of any union without a collective contract and with easy repression of activists. They trained generations of laborites in the art of leadership and taught common workers the value of organization. Not surprisingly, many of their leaders often became the leaders of the new unions during the early revolution. Pánfilo Méndez is a fine example. In 1907, he was vice president of Metepec's Círculo Fraternal de Obreros, signing the letter that founded the organization.[14] Less than a year later, he emerged as president of the Sociedad Mutualista de Ahorro, the de facto Río Blanco union.[15] He led the Orizaba workers' commission during the January 1912 negotiations that ended the general strike.[16] The workers' revolution sprang from these early activists and their years of struggle.

1910–1912: From Sociedades to Agrupaciónes de Resistencia

By the end of the Porfiriato, unionists had serious problems but also valuable experience. Nonetheless, the prevailing legal and power relations usually permitted owners to defeat strikes and labor organization. Because the premise of those defeats was state support for owners, when Madero mounted an effective challenge to Díaz, unionization changed radically. As the Labor Department wrote in 1913: "There is a very clear tendency among the workers to organize; they feel that in these groupings they will find the manner of reaching their goals . . . In these organizations, the leaders are not the most competent, educated, careful workers, but the most valiant, those that respect little their bosses and in order to get in good with their comrades, they promote strikes."[17]

Marjorie Ruth Clark added that "At once there began what amounted to a riot of labor organization."[18] Textile workers rapidly unionized the mills during the early revolution and never retreated from that unionization. It was a permanent revolution that lasted as long as the industry did.

Initially, "the law against such organization, to be sure, remained unchanged,"[19] so that other factors were more important, particularly the weakening of the repressive capacity of local government. The change began before Díaz left office, accelerated under de la Barra, and was in full swing when Madero became president. As one observer commented, "The Díaz ruthlessness was gone. Workers could meet, talk of organizing, and dream of striking successfully."[20] As repression weakened, the experience of labor activists and the strong desire for unions came to the fore.

The desire for labor organization is a constant theme in letters written by mill hands. In 1913 the mill hands in La Victoria wrote that without a union "we are completely backwards because the manager does whatever he wants."[21] In 1914, three Orizaba mill hands requested a "Union that we need so much.[22] In 1915, Río Blanco workers demanded recognition of the union and other labor rights "because they are just."[23] Textile workers believed that unions would improve their lives and fought to establish or sustain them. Labor's victory in the 1911 general strike strengthened that belief.

Building the unions was an important accomplishment because it was not automatic that the revolution itself would organize workers. To the contrary, violence hammered industry, made employment precarious, and did not substitute one set of recalcitrant owners for another more benign group. During the most difficult years, the battered economy cost jobs while inflation diminished living standards. As Alan Knight noted, armed revolution, economic collapse, and destitution affected urban labor.[24] During the contemporaneous Russian Revolution, economic collapse so devastated industry that autonomous workers' organizations suffered a death blow, which was not the case in Mexico.

Unionization in the early years was never easy because of the fierce opposition of owners, who often fired and blacklisted labor agitators and union leaders. In May 1912, El León fired Jesús Castro and Ismael Torres "for belonging to the union leadership."[25] The owners claimed that the two men led the workers "on strikes that have caused the ruin of workers and great prejudice to the industrialists."[26] In July, Elías Cano, a Mayorazgo worker, complained that he was fired "for belonging to the union leadership . . . who the administrator sees as his worst enemies, only because he can no long commit injustices with the workers."[27]

The risks for organizing unions were greater than just getting fired. The Díaz government sometimes sent labor activists to the army or to jail, a practice that continued under Madero, although somewhat reduced.[28] The Madero and Huerta governments diminished the harassment of labor activists but still viewed them with great suspicion because, as labor inspector Miguel Casas reported in 1913, "the workers associations have produced a lack of confidence and hatred among the industrialists, who oppose by all means and with full force their formation or existence."[29]

A key to labor organization during the early revolution were the labor activists willing to confront the dangers of leadership. The owners tried to eliminate these "dangerous agitators" because they believed that "labor discontent and the continuous strikes are caused by the leaders of the unions."[30] When the Miraflores workers went on strike in the aftermath of the 1912 convention, the owners complained about "Filiberto Dena . . . agitating the workers to join the strike of his comrades in the San Antonio Abad, Colmena and Carolina factories."[31] The owners of one mill warned the government that "if energetic measures are not taken against the workers, who are few, of bad will towards capital, they will lead us to dangerous upheavals and seriously prejudice these sources of wealth."[32]

José Otáñez was a typical "dangerous agitator of workers." The Labor Department warned local authorities that "his goal is to agitate in the Puebla factories, I beg you to treat this with energy by using all legal means to severely punish him."[33] Governor Meléndez promised that Otáñez would be "thrown out of the state,"[34] but the Labor Department insisted that he be "thrown in prison for some time."[35]

Such activists shared with rank and file the idea that workers were a humble and exploited social class who deserved the equality of citizenship. While some were motivated by formal ideologies, mostly anarchist, most activists did not come from formal political parties or anarchist groups. There was a widespread popular anarchism among many workers, if that is understood as antiauthority, antistate, and to a certain degree anticapitalist sentiment. Only to a lesser degree did formally established anarchist groups play a role in some early unions. The evidence suggests that such groups were not a major factor in the industry's unionization. Instead, a kind of popular workerism drove unionization in the early years, not unrelated to the popular challenge to authority that we saw earlier.

Along with labor activism, the early institutionalization of labor affairs that began with the 1912 settlement also spurred unionization. The agreements required labor representatives. If government was to use this protocontract to bring peace to the mills, it needed a legitimate workers' organization to represent workers, though finding such an organization was not easy. While there were many unions and proto-unions in a number of mills, there was no centralized leadership. The old Confederación Nacional de Obreros Esteban de Antuñano existed but was not a force among workers. Directed by Pascual Mendoza, it followed "a conciliatory position with authorities and owners."[36] In the second half of 1911, textile workers founded the Sociedad Cooperativa de Obreros Libres (SCOL). With headquarters in the old La Constancia mill, there were branches in Atlixco, San Martín Texmelucan, and Tlaxcala. The workers elected the notorious José Otañez, president, Rafael Silva, vice president, and Agustín Vara and Alfonso Reséndiz, secretaries.[37] The SCOL played a role in the 1911 strike.

Looking for alternate leadership, the government set up the CCOR in January/February 1912 as the labor counterpart of the Comité de Industriales. Textile workers now enjoyed two active national organizations, the SCOL and CCOR. As important as these national organizations were, the strength of the textile union movement was in local activists and factory unions, not national organizations. The local unions had been carrying out strikes and labor actions throughout 1909, 1910, and 1911. When labor inspectors fanned out across the country to settle the December 1911 strikes, and then the August/September 1912 strikes, they found in almost every mill "the union leadership that represents the workers in this factory."[38] No national organization controlled them.

The July 1912 convention gave the Comité Central Permanente de Obreros de Hilados y Tejidos de Algodón de la República the task of making sure "that the wage scale is being applied correctly, collaborating with this Department in favor of the workers progress."[39] The Comité had a headquarters in Mexico City and branches in the various textile centers. Claiming that it "officially represents" workers, the Labor Department gave it a budget of 200 pesos a month.[40] It was expected to function as "a mixed arbitration tribunal designed to make decisions on the difficulties that arise with the application of articles 5 and 12 of the Factory Rules."[41] For the next two years, the committee intervened in various disputes.[42] Thus, labor organization was built into the fabric of the 1912 agreement.

Many workers did not like the Comité, particularly the anarchists who created their own national labor organization in July 1912, the Casa del Obrero Mundial (COM).[43] Though not a union, the Casa quickly became a leading labor organization of the early revolution, pursuing broadly anarcho-syndicalist goals. Some of the anarchists had previous labor experience in Mexico, particularly through the Confederación Tipográfico de México, which became the Confederación Nacional de Artes Graficas.[44] Juan Francisco Moncaleano, Colombian anarchist and fugitive recently arrived from Havana, started publishing *Luz, Periódico Obrero Libertario*. The organization played an influential role in the labor movement during the early revolution, and prominent members included Luis N. Morones, Celestino Gasca, Salvador Alvarez, Samuel Yúdico and Eduardo Moneda, who later became the founders and leaders of the CROM, as well as Antonio Díaz Soto y Gama and Rafael Pérez Taylor, who later joined the Zapatistas.[45]

Not surprisingly, the Casa hated the Comité Central. In March 1914, its newspaper, *El Sindicalista*, argued that "official institutions corrupt the efforts of workers," adding that "The five members of the Comité, have fully demonstrated their ineptitude and bad faith to resolve problems and to defend the rights that have been entrusted to them."[46] It demanded that workers "refuse to recognize the Committee and consider it among their great enemies."[47] Despite this advice, most textile workers did not disown the Comité because they believed it strengthened unionization in the mills. When the COM changed its political line in order to support Carranza against Villa, there is some evidence that it changed its position on the Comité, trying to gain control rather than destroy it.[48] As always, its influence on textile unions was limited.

Mill hands were probably right in believing that the Comité spurred unionization. Receiving and mediating workers' complaints, it reinforced the idea that labor organizations helped working lives. An example was the old conflict between maestros and workers. Before the revolution, workers could only complain to the owners about their immediate bosses, but the complaints usually fell on deaf ears. The Comité changed that by legitimating their complaints and to some degree protecting the complainers.[49] It interpreted the new work rules and wage scale in ways sometimes favorable to workers. It allowed for worker input and showed workers that their organizations could influence work rules.

The new Labor Office also strengthened unionization. From de la Barra to Huerta, the Office was not an enemy to unions, though not always their friend. Its primary goal was peace in the factories, often understood as satisfying the owners. Nonetheless, its very existence provided unions a forum and protection, however weak, which activists used. When federal labor inspector Miguel Casas traveled through Puebla and Tlaxcala in 1913, he asked each local jefe político for a room in which to lecture workers. He requested that factories send workers to attend the meetings, at which he lectured on the 1912 agreements and "good doctrines on economy, order, and morality." He reported that "I had the great satisfaction that a large number of workers learned perfectly well to manage the wage scales, and absorbed the lesson of the necessity of avoiding strikes and disorder."[50] He visited local factories with the same intention. It was impossible for workers not to gain confidence from such visits. When he returned to Mexico City, Casas had revised thirty-eight wage lists, visited nine factories, and held twenty conferences with workers.

After 1910, forming unions was therefore not a moment but a process. The desires of workers, the activities of militants, and the new institutionalization all played important roles. Between November 1912 and February 1913, another wave of textile strikes spread across the country, many of which pursued contract enforcement and defense of unions. Coming on the heels of the 1911 general strike and July 1912 accords, it further strengthened unionization.[51]

The unions during this period had relatively simple internal structures. At the initial meeting, the members elected a leadership, the *mesa directiva*, which usually included a president, vice-president, and one or two secretaries. Workers, managers, and government officials referred without distinction to the leadership (mesa directiva) or the union (sociedad). Most still did not use the term *sindicato*. For example, Antonio de Zamacona, in commenting on the union at El León, wrote about "*los obreros Sres. Jesús Castro, Ismael Torres y Eduardo M González que forman parte de la Mesa Directiva de una Sociedad que tienen constituida el gremio obrero de la fábrica.*"[52]

Leaders generally represented members and did not rule over them. The evidence suggests that workers thought of themselves as rulers of the unions. When the El León union sent representatives to Mexico City in May 1912, to explain the recent strikes, the letter spoke of two workers "commissioned by the Workers of El León Factory," rather than representatives of the union.[53] Government officials and owners certainly believed that leadership represented the workers, if not their interests, often accusing them of deceiving workers. They also, on occasion, accused leaders of corruption. Labor Inspector Zamacona claimed that the Miraflores union deceived the workers "in order to get a weekly contribution of 10, 15, and even 20 centavos."[54] They also accused male union leaders of deceiving female union members because they could not accept that women were militant unionists. William O. Jenkins, owner of La Corona, reported that "union officials . . . have deceived some of the girls in this factory, making them believe that they are unhappy."[55] Making them believe they were unhappy! In fact, women as well men lost no opportunity to form unions because they were in fact unhappy. Not all workers joined unions, but those who did seemed to support their union leaders, all of whom were workers in the mills.

1913–1916: Revolutionary Consolidation

Because of the dramatic events of the early revolution, as early as 1913, the position of unions was quite different from only three years before. Outright and effective repression by a strong central government, entrenched local governments, and powerful owners had ended. The federal government had taken notice of industrial workers and created a labor department. There was an official industrywide agreement in cotton textiles, with federal inspectors spread out across the country to enforce it. Trade unions had organized the largest mills and now played a role in an emerging labor relations system, often with support from government officials. It was not a perfect system but to workers it seemed better than the old one. For example, when El León tried to reduce piecework rates in early 1913, the unionized workers walked out, the Labor Office intervened, and the mill restored the old rates.[56] Workers supported unions because of their real impact on their lives. The Labor Department also believed that the new system avoided "many conflicts between owners and workers."[57]

The owners, however, still could not come to terms with this new reality. When Huerta came to power, many hoped that he would return the Pax Porfiriana, but in labor affairs they were mistaken. Adalberto A. Esteva took over the Labor Office Director and ordered the mills to cease harassment of unions and labor leaders.[58] When the members of the Comité Central Permanente resigned, he convened a meeting of textile workers on May 26, 1913, to name a new Comité.[59] That same month the Labor Office sent a letter to the governors asking them to establish Cámaras del Trabajo in their states.[60] Even this modest proposal, which allowed owners and politicians to outnumber labor representatives, provoked the ire of some governors. The governor of San Luis Potosí wrote that the Cámara would "awaken insane ambitions . . . bringing more difficulties because of unjust demands"[61] Meanwhile, Huerta proposed elevating the Labor Office to ministerial status, changing the Secretaría de Industria y Comercio to the Secretaría de Industria, Comercio y Trabajo (SICT) and absorbing the Labor Office.[62] He noted that "the labor question moves all governments and all peoples today with greater force than ever: it is the contemporary problem, the problem of the century."[63]

By 1914, textile workers had seen three successive governments— de la Barra, Madero, and Huerta— cave in to militant unions. While labor militants and labor leaders acquired new status, some factory owners and local political bosses continued to fire and blacklist the leaders. The leaders, with worker support, fought back. When a Guadalajara owner fired Bernabé Nuñez for organizing strikes in Juanacatlan, Atemajac, and La Experiencia, Nuñez tried to kill him, firing at least three shots. Afterwards, Nuñez continued to lead strikes and later threatened a federal labor inspector.[64] Activists in El Carmen (Atlixco) stoned the scabs in El Pilar for trying to break the union.[65]

Meanwhile, incipient institutionalization brought internal changes to unions. More workers began to pay regular union dues, which in turn increased possibilities for corruption.[66] On November 10, 1913, Andrés Cabrera, vice president of the San Félix union (Atotonilco, Puebla) accused Ysidro Flores and Manuel Lembrino of trying to "destroy our union," by spending union funds without informing members.[67] Flores and Lembrino got the local judge, Juan Cedillo, to assemble thirty union members at the courthouse to investigate.[68] The testimonies revealed that the leaders had been collecting ten centavos a week from the factory administration for each worker, though they claimed to not know the funds on hand or recent expenses.[69] The judge estimated that from January to November, the eighty-member union had received 291.12 pesos, and spent 246.28 pesos. A short meeting followed in which members agreed to dissolve the mesa directiva, to divide the remaining funds among themselves, and to stop collecting the 10 centavos weekly.[70] Cedillo asked the Labor Office to help form a new mesa directive. Mexico City answered that "serious difficulties prevent the Labor Department from sending a delegate," adding that the Office was studying "new forms of regulating and establishing statutes for workers associations."[71] Two weeks later, mill hands met again with Judge Cedillo, elected new leadership, and once again authorized the new leaders to collect 10 centavos a week from each worker, with new rules of financial accountability. Lembrino, one of the two original dissidents, was elected union secretary. This remarkable event in late 1913, at the height of civil war against the Huerta government, sheds much light on unions at the time. They had become important enough to collect and spend dues. Despite corruption, rank and file supported their union. Furthermore, rank and file had the political strength to remove corrupt leaders.

The years 1914 and 1915 were brutal, as violence and revolution swept across the country. The United States invaded Veracruz, forcing Huerta to flee the country. This left Carranza and Obregon free to attack Villa and Zapata. For a short time there were two governments, one in Mexico City and another in Veracruz. Business and the economy suffered greatly. In the textile industry, the Labor Department reported that many mills could not get cotton and that "their products lack an outlet."[72]

Without doubt, this ongoing violence transformed unions and the workers' revolution. First, it forced competing elites to appeal to workers, giving mill hands much greater power than they would have had otherwise. Second, the years of continuous violence and civil war taught workers that power indeed came from the barrel of a gun. As unions became more institutionalized and more powerful, disputes for control over unions would be settled the way other disputes were settled during the revolution, with guns. Under the conditions of revolution, in Mexico it became clear that those who organized killing best would gain control over mass organizations like unions.

When Huerta fled the country, he took with him the reaction, leaving only various factions of revolutionaries to compete for power. All of them sought to win favor among industrial workers. Even the most conservative, Carranza, supported unionization. In December 1914, his Labor Department argued "it is urgent . . . that the right of unionization be made effective."[73] When Carranza extended the life of the 1912 textile agreements in December 1914, he mandated that "textile workers will elect union leaders in each factory who will represent them in case of violations of the Wage Scale and Work Rules."[74] His Labor Department Director Marcos López Jiménez met with San Lorenzo workers in January 1915 to tell them that "the object of the visit, was to install a union [mesa directiva] and we then formed the union [*agrupación de Resistencia*]."[75] He met with workers of the other Orizaba mills, "for a union that will protect the interests of the workers."[76]

When the Santa Rosa workers formed their union, they immediately demanded overtime pay as mandated by the new military decree, taking their complaint to the recently created Tribunal de Arbitraje. The mill's owner, Camilo Maure, could not ignore the demand because the Tribunal forced his presence at a hearing. In May 1915, the women seamstresses in La Suiza formed a union and in July rejected a 10 percent wage hike as insufficient, turning to the Labor Department to help them get a requested 25 percent.[77] The rising institutionalization contributed to unions.

López Jiménez also participated in the creation of the Agrupación de Resistencia, a regional organization of the local unions headed by Enrique Hinojosa.[78] The Agrupación then sent Vicente Cortez to meet with the Puebla unions to urge their support for the Labor Department. The meeting led to the formation of the Agrupaciones de Resistencia de Puebla, led by Agustín Rosete.[79] The formation of regional federations in Orizaba and Puebla made local unions more powerful. As the leader of the Cerritos union noted in 1915, "we have formed unions and we have seen since then that they pay more attention to us; it seems that the stimulus to organize has made the industrialists more inclined to cede a bit more."[80]

Thus unions continued to develop even while the violence of the broader revolution reached new heights. The textile zone from Mexico City through Puebla to Orizaba was at the very center of the fighting. Zapatistas periodically raided Atlixco and sometimes Mexico City. Veracruz experienced revolutionary violence and the invasion by the U.S. military in 1914. Two examples from Labor Department files illustrate the daily impact of violence. When Labor Inspector Daniel Galindo tried to travel from San Félix to Texmelucan in November 1914, "we found out that the garrison stationed in El Molino, rebelled." When they tried to go to San Félix instead, the rebels "had taken possession around the factories, not letting anybody leave." On the return to San Martín, Zapatistas surrounded the town, "[so] we couldn't continue our trip until 9 pm when the Constitutionalists had cleared out the Zapatista forces."[81] When Labor Inspector Miguel Casas visited textile factories in San Angel and Contreras, to the south of Mexico City, in July 1914, he found that La Hormiga worked a shortened schedule because of "the shooting that there has been."[82] Zapatistas occupied Contreras with its mills, Casas reporting that "the towns of Contreras and La Magdalena are entirely destroyed by fire and only one or another house remain."[83]

It is difficult to calculate the effect of violence on the thoughts and feelings of common workers but it must have been considerable. What we do know is that workers themselves increased labor violence to get their way in the mills. Mill hands sometimes shot, stabbed, clubbed, burned, and beat owners, supervisors, authorities, and scabs. Why not? By 1915, there was no political authority in Mexico other than violence.

Why would unions be different? Violence in the mills became accepted process. Until 1915, however, the political violence of unionized workers was mostly directed against bosses. What would happen when the unions confronted internal problems?

One group that continued to spur unionization and espoused violence was the Casa del Obrero Mundial. Its newspaper, *El Sindicalista*, argued that "The citizen of modern societies is a slave of the land, factory or machine; he suffers hunger and humiliation, is exposed to misery, the hospital or the asylum; suffers the insult of the owner or the slap of the supervisor."[84] During the unionization drive of early 1915, Casa propagandists went from factory to factory telling workers that the Labor Department, which had been organizing unions, "is useless and cannot solve anything in your favor . . . ; in order for the Orizaba workers to get what they desire, it has to be done by bullets and not by laws."[85]

According to Clark:

Wherever territory was conquered by Carranza's forces, there the labor groups established branches of the Casa del Obrero Mundial . . . The method of procedure was well defined. Representatives of the Casa arrived in a city and, utilizing official support and any local incipient labor organization, organized a branch of the Casa del Obrero Mundial. This accomplished, they called a strike or strikes, usually for higher wages, recognition of the union or unions involved, and the eight hour day. If the strikers won—and since they had the support of the military commander of the district they could scarcely fail to win—the newly created branch of the Casa became at once the center of the organized workers.[86]

In Mexico City, the COM was successful in organizing workers in urban transport, electrical power, and the telephone and telegraph company. The latter carried out Casa-led strikes in late 1914.[87] The government's defeat of the COM's general strike in 1916 led to its quick demise, after which the CROM replaced it as the leading force in Mexican unionization.[88] In its glory years, however, the organization spurred the hopes of many industrial workers. During that period, its politics moved from militantly antigovernment to a later alliance with Carranza against Villa and Zapata.[89] When Carranza finally turned on the COM, it was finished. In textiles, however, unionization and the radical activities of textile workers remained unaffected by the demise of the COM, except in a few local cases.

Gruening wrote that "Much of the past went by the board automatically in the Revolutionary years 1914–1916. Physical abuse of workers,

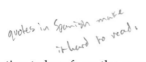

peonage, the obligation to buy from the company store ceased, and the
right to organize was combated less openly, but none the less bitterly
for it was recognized by both sides as the preliminary to further demands by the workers."[90] By 1916 the position of unions had become
more stable and more regularized. In some states, new military decrees
and state labor offices defended unions. Most unions were democratic
and voluntary, unaffiliated to larger organizations. They were rowdy,
engaged in numerous labor actions, and as violent as the revolution
that surrounded them.

1917–1920: Organized Labor Emerges

By late 1916, the unionization of the industry was essentially complete.
In February 1917, Article 123 presented trade unions with an enormous
legal victory. Clause XVI declared that "Both employers and workers
shall have the right to organize for the defense of their respective interests, by forming unions, professional associations, etc."[91] Clause XVII
added that "The laws shall recognize strikes and lockouts as rights of
workmen and employers." Workers had fought for this since the Porfiriato and now they had won. The new legalization did not spur
unionization, which had already taken place, but it did generate institutional change because previously informal *agrupaciones* now became
formal and legal *sindicatos,* which in the beginning meant registering
with local government.[92]

When Carranza crushed the Casa-led general strike in July 1916,
Álvaro Obregón intervened to save some of the strike leaders, and
many of them later repaid the favor by supporting Obregón, especially
Luis N. Morones. Before its demise, the Casa had organized the Federación de Sindicatos del Distrito Federal in 1914.[93] It was of little sig-
nificance until Morones, head of the electricians union, became general
secretary in 1916, helping to revive the organization. Under his leadership, the January 1916 Declaración de Principios used both the old and
new terms *sindicato* and *agrupación,* stating *"Los sindicatos pertenecientes
a la Federación, son agrupaciones exclusivamente de resistencia."*[94] In March
1916, the Federación organized a national labor congress in Veracruz.
Although a well-known local anarchist, Hernán Proal presided over
the meeting, Morones led the unionist group that bitterly opposed the
anarchists.[95] Morones believed that trade unions tied to government
would yield more for workers than the direct action proposed by the
anarchists.[96] A subsequent national labor congress was held in Tampico

in October 1917, originally convened by local stevedores and the local COM. The congress met to discuss national labor unification.[97] At the conference, Morones's supporters, a slight majority, again fought with the anarchosyndicalists.[98]

After the Tampico conference, Governor Gustavo Espinosa Mireles, an ally of Carranza, appointed an organizing committee to plan for a third conference.[99] By now it was clear that the COM would not recover its former influence and that the various groups allied under the Constitutionalist banner would form the nucleus of a postrevolutionary state. Hoping to influence the formation of a cooperative labor central, in May 1918, the government of Coahuila hosted a labor congress in Saltillo. One hundred sixteen organizations from eighteen states sent representatives to the Congress, which named an executive committee of Morones, Jacinto Huitrón, Teodoro Ramírez, and Ricardo Treviño. The Congress founded the CROM. Morones, Treviño, and J. Marcos Tristán comprised the first executive committee. The CROM took as its theme "Salud y Revolución Social."[100] Huitrón and the anarchists quickly left the organization. The powerful Federacion de Sindicatos del Distrito Federal also stayed out. Meanwhile, Morones became secretary general and soon established an alliance with Samuel Gompers and the American Federation of Labor.[101]

Morones quickly created Grupo Acción, a secret and disciplined group with great personal loyalty to Morones.[102] Morones and Acción ran Mexican labor affairs in the 1920s.[103] Morones believed that the Casa had derived strength from its alliance with the state and collapsed when it challenged government. Having learned from and then combated anarchists in the labor movement, he thought that workers had more to gain from a state-labor alliance than from an independent stance. In the unstable social and political climate of late revolutionary Mexico, the CROM quickly became one of the country's most powerful political actors.

In August, 1919, Morones signed a secret pact with Carranza's former military leader and now political rival, Alvaro Obregón. He promised to support Obregón's presidential ambitions while Obregón assured the CROM privileged access to government positions. Many of the anarchists who supported the CROM to that point then quit to later form the rival CGT. Morones meanwhile founded the political wing of the CROM, the Partido Laborista Mexicano (PLM), in August, 1919. He used the PLM to support the electoral ambitions of his allies. When Obregón assassinated Carranza in 1920, his alliance with

Morones and the PLM assured the urban electoral support with which he won election to the presidency later that year.

Meanwhile, in Atlixco, Puebla, and Orizaba local labor leaders began to affiliate their unions to the emerging CROM. Pánfilo Méndez, early labor organizer, had joined the Casa.[104] Former members of the Casa subsequently participated in establishing the Union de Resistencia del Ramo Textil (URRT) in 1917. In Puebla, the URRT became the Confederación Sindicalista del Estado de Puebla (CSEP).[105] Ignacio Salazar and Onofre Armijo of the Sindicalista attended the 1918 Saltillo convention that founded the CROM. Upon returning to Puebla, they affiliated the Sindicalista to the new confederation, bringing in the unions previously affiliated to the now disappearing Confederación Nacional de Obreros Esteban de Antuñano.[106] In 1919, unionists established the Federacion Sindicalista de Obreros y Campesinos de Atlixco (FSOCDA) as an affiliate of the CSEP; its first secretary general was Baraquiel Márquez.[107] Like the national CROM, the Sindicalista established small groups of local leaders to manage union affairs, Alpha in Puebla city, Económico in Atlixo.[108] Although the Sindicalista was mostly a textile organization, it also organized workers in other industries. In Puebla, it maintained a certain distance from the CROM´s national leadership, while the leadership in Atlixco was independent of the organization in Puebla City. Nonetheless, they all shared basic ideas, strategy, and ties to national labor leader Luis N. Morones.

The early CROM may have been dominated by Morones but many of its founding members had come from the COM and had been schooled in anarchosyndicalism. During the organization's 1920 convention, the communists, anarchists, and some syndicalists attempted to defeat Morones. When they failed, they left the organization to found the CGT, led initially by Rafael Quintero, Rosendo Salazar, and José G. Escobedo.[109] By the end of 1921, the Communists withdrew from the CGT, leaving it to the anarchists.[110]

With the founding of the CGT, there were now two national labor organizations active in the textile centers of Atlixco, Puebla, and Orizaba. Pablo Rueda and Baráquiel Márquez, formerly of the CROM and the Federación Sindicalista de Atlixco, participated in establishing the CGT, immediately giving it an important presence in Atlixco. Márquez then created the Federacion Local de Obreros y Campesinos (FLOCCGT) to replace the CROM Sindicalista.[111] This made anarchosyndicalism a powerful force in some of the factories, particularly El Volcán and El Carmen.[112] In 1919, radical elements created the Federación de

Sindicatos de Trabajadores de Hilados y Tejidos del Distrito Federal, Estado de México y anexos, an important labor center among the capital's textile workers. It participated in the establishment of the CGT.[113] With competing regional and national organizations, the CROM and the CGT would battle for control of the textile industry.

What made these labor wars so significant was that unions had achieved an importance that even the most militant activists of 1910 could not have imagined. The combination of powerful union confederations tied to the state plus the new legal environment meant that controlling a union in 1920 was not at all like leading a union in 1910. In 1910, being a labor leader required great heroism and self-sacrifice. Many were fired and lost their jobs, homes, and possessions. By 1920, heading a union gave the leader a number of highly lucrative opportunities, from controlling jobs to union dues to opportunities in local and national politics. Without doubt, it was the victory of the Constitutionalists and Article 123 that spurred the formation of long-lasting national labor confederations. The Constitutionalists were not a united group but rather a loose alliance of regional warlords. They needed allies to cement their victory. Radicalized workers represented a political opportunity for them. Meanwhile, the new legal environment provided unions unprecedented protections and power, but it made the victories of the workers' revolution dependent on the institution from which law emanated: the state. A union-led revolution had to turn to the state, and the new postrevolutionary state turned to the unions. Each made the other more powerful, which is why these natural allies ruled Mexico in the 1920s and 1930s.

CHAPTER 8

Labor Conflict in the Early Institutional Period, 1917–1923

BETWEEN 1910 AND 1917, revolutionary cotton textile workers achieved two institutional breakthroughs: proworker labor law and strong unions. The workers' revolution, however, did not end with Article 123 or with the establishment of a formal labor relations system. The structure of the informal labor relations system, the unwritten social rules that governed work and the social relationships of work, remained to be determined. This chapter looks at three areas in which the workers' revolution moved forward during the early institutional period from 1917 to 1923: labor violence, trade unions, and the shop floor. Workers and unions continued their revolution, albeit not unchanged, during an unsettled period in Mexico's revolutionary history.

The country remained deeply conflictive in the years following the 1917 Constitution. The Constitutionalists did not liquidate Emiliano Zapata until 1919 and Francisco Villa until 1923, both falling to assassins' bullets. They managed to elect Carranza president in March 1917, but he suffered an unsuccessful assassination attempt in April 1920 and a successful one in May. Alvaro Obregón then became president, though he too would fall to the assassin in 1928. These were only the most famous victims of early postrevolutionary violence, as thousands of other Mexicans also died in the fighting and infighting that characterized the public sphere. On the other hand, the Constitutionalists had effectively won the revolution and civil war; they enjoyed a new Constitution, and economic recovery was relatively rapid.

If high politics was unstable in the extreme so was the labor regime. The new Constitution and subsequent state labor codes ushered in a new formal labor regime but the informal rules were yet to be established. Into this vacuum rushed the old violence because that is what all parties had learned during the previous period. Mill hands in particular were keenly aware that they had won their new unions and laws through conflict rather than cooperation. Negotiations only took place after walkouts, strikes, assaults, and murders. They watched as elites settled conflicts through murder and assassinations. With continued violence in the country after 1917, who would rein in the workers if they continued to employ the old tactics? Thus, although overall violence diminished somewhat, the opposite took place in cotton textiles. The first section of this chapter explores the changing nature of labor violence in cotton textiles during the early institutional period.

The same factors that contributed to a change in the nature of labor violence also created two important shifts in trade unions: a turn to regional and national federations and a realignment of the relationships between union leaders and rank-and-file workers. The second section of this chapter looks at how these changes came about and with what effect.

Finally, it is important to note that from the beginning cotton textile workers sought two basic goals: a better standard of living and more worker control over the shop floor. The third section of this chapter shows that workers continued to push for both of these during the early institutional period, suggesting that the workers' revolution was alive and well, albeit modified, until 1923.

This chapter and book conclude in 1923 not because the workers' revolution ended there, which it didn't, but because the changes mentioned above signaled a new period for industrial workers, that of powerful trade union confederations tied to government. The preliminary data suggest that the workers' revolution did not suffer a defeat in the 1920s, but that it changed so radically that the period deserves a study of its own, one that would include the labor wars of the 1920s, a new industrywide contract in 1927, the incorporation of labor confederations in a new ruling party, the rise to power of local labor leaders, and an as yet unstudied change in the power and social relations inside factory unions. However, while we may not fully understand the workers' revolution in the 1920s and 1930s, we do know what happened between 1917 and 1923, the subject of this chapter.

Violence and Instability

The violence of a broader revolution had legitimated the labor violence of the workers' revolution. Nonetheless, as large-scale warfare diminished after 1917, in the mills and the mill towns, the labor violence escalated. Following a regional strike in 1918, fearful Puebla owners requested police presence in their mills.[1] In 1921, the CIM's Sánchez Gavito reported that, "In the factories . . . workers are entering work with armed objects including pistols . . . For this reason, administrative personnel have reported to their superiors that they cannot continue working in that manner, because they are not resigned to losing their lives, at any moment, for carrying out their job of command and order within the factories."[2] More and more, the mills became armed camps.

The violence ranged from minor property attacks to physical assaults and murder. The petty attacks were generally a means to an end, as when in late 1918 workers in the Santa Rosa mill tore up the cloth to protest a quality change.[3] Unionists also began to enforce internal discipline through threats or violence so that when Remedios Aguilar decided to work two shifts in order to earn an extra income, the other mill hands warned him that "se atendiera a las consecuencias," meaning they would kill him if he continued his recalcitrant ways.[4]

After 1920, the conflicts became much more violent. Early that year, intraunion fighting in Orizaba led to expulsions. Some of the workers on the losing side emigrated to the mill at San Juan Amatlán, led by Pedro Sosa, a former member of the successful 1916 worker commission to Mexico City.[5] The owners called them "professional agitators who redoubled their perverse labor."[6] Despite an agreement to not hold union meetings on company time, the mill hands took a Friday afternoon to discuss labor matters, angering administration. The next day, when the second shift entered the mill, chaos erupted as workers began whistling, shouting, and banging on the machines, forcing the factory administrator to send one of the employees to restore order. Instead of order, a group of eight or ten workers grabbed the *empleado* and attempted to kill him with a long knife when the administrator, armed with a pistol, rushed in. He wounded one of the assailants, taking a knife in the arm while doing so. He ran off to grab a rifle, climbed to the roof of the mill, and began shooting below, while taking the time to extract a knife from the back of another employee who had joined him on the roof. The mill hands fled the building, but not without starting a fire under a couple of looms.[7]

The police arrested some of the workers, leading to a demonstration in town during which protestors shouted "Long Live Free Russia," "Death to government," "Long live the World Revolution," and "Death to the Spaniards." Confronted by the crowd, municipal authorities quickly released the prisoners and instead jailed the factory administrator. The alarmed owners of the mill wrote to the state's governor to tell him that:

it is an extremely grave matter, because it is a question not only of more or less absurd petitions against capital, nor of strikes thinly veiled with legality, but an eloquent demonstration of Bolshevism, which, with the incendiary's wick in one hand, and the fist of a murderer in the other, they attack Public Power, trampling without any scruples, the principle of authority, the only basis on which the state can conserve itself.[8]

As if to underscore their fears, a few days later, somebody tried to kill the mill owners as their car sped across a bridge, the shots narrowly missing their target. Shortly thereafter, the workers recently released from jail returned to the mill, with the goal of organizing nearby mills and perhaps provoking a general strike.[9]

The violence in San Juan Amatlán spread to other mills and extended the strike movement to the entire state of Hidalgo. In Mexico State, striking workers violently prevented scabs from entering the mills. In La Colmena, after blowing up targets with dynamite, workers occupied the mill.[10] During a later conflict in Atlixco, the municipal president feared that "the workers united with the campesinos" would assault the municipal palace, "exercising violence against the Authorities."[11]

In the fight between opposing union groups, Atlixco became especially violent. In October 1921, a group of unionists threatened to kill opposing workers if they entered the Metepec mill.[12] The next month brought death threats against the mill's administrator, Constantino Matilla.[13] In December, unionized workers, armed with guns and other weapons, attacked a group of scabs headed to nearby El León, forcing them to retreat.[14] Such violence against scabs was the norm and, as one noted, "there has been blood spread."[15] It was understandable that a few months later, Bernardo Barrolo, the mill's administrator, requested permission to arm factory guards "inside the factory with 30-30 carbines each one."[16] Scarcely a month later, the guards heard shots around midnight just outside the factory door. They rushed to "repel the aggression." One of them saw somebody try to jump the mill's south wall but drove him off with a few shots, after which another three or four men jumped the wall.[17]

The violence continued unabated between 1920 and 1924, during which a police report listed the highlights:[18]

1. November 10, 1921, Fidel Luna shot Juan Montiel.
2. Carnaval, 1922, Justo Vázquez, *libre*, murdered Pedro Flores.

3. March 19, 1923, Sixto Romero, *libre*, wounded Margarito Fuentes.

4. April 5, 1923, Matilla helped a group of Metepec *libres* attack the CROM members at El León, where they killed five and wounded another.

5. April 7, 1923, a group of *libres* attacked Julio Vargas.

6. April 15, 1923, Baltazar Pérez wounded Sotero Pavón.

7. April 29, 1923, six *libres* attacked Salvador Paredes, while Pedro Gutierrez and Rodolfo Torres beat up Guadalupe Martínez Moron and Jose de Jesús Soto.

8. May 1, 1923, *libres* attacked Metepec workers when they returned from the May Day parade and attempted to kill the union's Secretario General, Alejandro Haro.

9. June 4, 1923, Cromistas alleged that Matilla ordered two *libres*, Salvador Paredes and Luis González, to assassinate the Secretary General of the CROM union "in order to finally end this organization in the factory."

10. September 23, 1923, the head of the CROM union was wounded for a third time, this time by Esteban García, who fled town.

11. December 24, 1923, in the midst of de la Huerta rebellion, Pedro Ledesma, Fidel Luna, Pedro Gutiérrez, and Vicente Armenta attacked and wounded Ramón Flores, went to his house and, after stealing some money, kicked his pregnant wife, who moments afterward, "had a failed pregnancy."

San Juan Amatlán and Atlixco were not atypical examples of an extremely complex situation in the cotton textile industry after 1920 where four interrelated processes drove labor violence. First, the lack of clarity about the new institutionalization of labor affairs drove owners to test the system. The owners believed that the new unions, labor laws, and government offices infringed on the natural rights of ownership. They begged the federal government to spell out the exact role of unions.[19] The head of the Chamber of Industrialists of Orizaba sent a letter in 1918 to the Veracruz State Labor Board that accurately describes their views.

A worker by writing or in person goes to the [Labor] Department Head, presenting a complaint against the owners. The Head of the Department orders the owner to go the Labor Department. In a meeting of the owner and the worker, the worker presents his complaint and the owner his evidence; on occasion, both parties are required to present proofs within a certain period. Given the evidence by both parties, the Department Head or sometimes the Municipal President emits a resolution which systematically favors the workers. The judgment must be carried out of course.[20]

The owners thus tried to resist the new institutions. They sometime used the old tactics such as firing union activists. The administrator of El Carmen fired several workers for collecting union dues inside the factory,[21] and Metepec locked out the workers to break the union.[22] After 1920, many of the owners shifted tactics when they realized that it was impossible to suppress unions. Instead, they tried to create company unions, organizing groups of so-called free workers, or *libres*. The problem was that this introduced a new group of violent unionists into the mix, resulting in the murder and mayhem in Atlixco that was mentioned previously.

Second, workers and unions also sensed a lack of clarity, so they tested the new institutionalization, which meant violating factory rules. Thus they tore cloth, damaged machines, carried out meetings during work time, or refused to obey supervisors. When El León locked out its workers in June 1919, the union appealed to Municipal President Eduardo Vivanco. Vivanco forced the mill to end the lock out, to recognize the union leadership, to allow the union to collect dues at work, and to allow union representatives to leave work to carry out union business.[23] After Vivanco left office, a municipal administration who was less favorable to the unions took office, but the new municipal president found himself outflanked when the local army battalion supported the union, which he reported when three hundred striking El León workers attacked the local jail under the passive eyes of federal troops, so that the town authorities could do nothing. The Mayor noted that the commander of the federal troops even provided guns to the unionists.[24]

When challenging the system led to inevitable reprisals by the administration, workers often responded with violence, as seen in the cases of El León and Amatlán. For both workers and owners, establishing a new formal labor relations system through military decrees, Article 123, and state labor codes, created an untested terrain of a new informal labor relations system. It was inevitable that workers would test the new system through labor violence.

Third, the new institutionalization ratified the power of unions, which transformed them. With unions now worth fighting over, it was also inevitable that competing groups would emerge to run them. The workers who formed these groups employed the methods they had learned during the struggle against owners to take control of the unions, the old violence. The bloodiest fights in this period were those between organized groups of killers at the service of the CROM, the CGT, or the Libres. As the Metepec administrator wrote in 1922, most

TABLE 8.1

Registered Textile Strikes and Strikers, 1920–1923

Year	Number of strikes	Number of workers affected	Number of workers on strike
1920	99	43,501	37,937
1921	106	42,235	37,799
1922	135	40,866	40,383
1923	73	40,139	36,253

SOURCE: Moisés de la Peña, La Industria Textil en México, El Problema Obrero y Los Problemas Económicos (Mexico, 1934), 28.

conflict came about "due to differences of opinion among the workers."[25] Of course, workers settled these differences of opinion with fists, knives, and guns, and sometimes even dynamite.

Finally, the evidence suggests that a powerful component of the old antiauthority revolt drove some of the violence of this period, with workers less restrained in their attacks on foremen, administrators, and owners. For example, when workers demanded the removal of a supervisor in El Mayorazgo, they physically assaulted him with pieces of iron as he tried to walk through the mill, forcing him to retreat from the shop floor.[26]

This combination—owners and workers testing the limits of a new system, competing union groups, the antiauthority sentiment of workers, the learned violence of a social revolution, and a still unstable state—makes it not at all surprising that labor violence escalated in the half dozen years following the passage of the Constitution.

The consequence, of course, was that cotton textiles remained a volatile, indeed explosive industry, as reflected in the strike data, 1920–23. In December 1921 and January 1922, federal Labor Inspector Roberto Saviñón visited twenty-nine textile factories in the Puebla/Tlaxcala valley. He found strikes in ten of the twenty-nine mills.[27] As the owners stated in 1922, their most serious problem was "la cuestión obrera." They argued that the Mexican worker was excellent until he unionized, at which point "it is impossible to deal with him."[28]

The New Trade Unions

Of course, most textile workers had unionized by 1917, the year in which the strongest unions pressured the federal government to hold a new meeting of the industry to update the 1912 agreements. Carranza held preliminary meetings but became fearful of increasingly powerful labor organization, so he suspended the gatherings, asking the governors to convene local meetings of industrialists and union leaders. Thwarted in their attempt to obtain a national wage hike, the unions launched strikes across the country. The governor of Mexico City ended the strike on May 12 with pay increases of 60 percent to 75 percent. On May 4, the Orizaba textile unions walked out. Their strike ended on May 17 with pay raises of 70 percent to 80 percent. The Puebla unions did not walk out, instead accepting an offer from the government and the industrialists to recognize their regional organization, the Unión de Resistencia. These coordinated regional strikes, facilitated by regional alliances of unions, demonstrated the power of regional and national labor federations. [29]

On July 19, 1917, Carranza decreed the free importation of cotton textiles, ending protection of the industry. In September, the industrialists shut down the mills. In response, Carranza seized the factories. He then introduced a new law to legalize the seizures. The industrialists opposed the law but Congress approved it on October 27. In order to avoid more government takeovers, the owners reopened the mills.[30] Carranza, however, was unprepared to have the state run the industry, so he eventually repealed the free trade decrees.

The following year, mill hands in El León and in the five Cholula mills walked out, demanding wage increases of 150 percent. Between March 5 and June 10, the Federación de Sindicatos de Puebla led the state's other mills out on strike. The governor and the state's Junta Central de Arbitraje called a meeting of labor leaders and mill owners, where the owners offered a 50 percent wage hike. They complained that their workers were not "the workers of before, that produced so much, today they are only dedicated to dealing with union matters and don't want to work."[31] The Federación rejected the owners' offer, so the strike spread to Tlaxcala and Orizaba, with perhaps 15,000 out of work.[32] The state legislature responded with a law to implement profit sharing but the owners obtained an injunction which the Supreme Court ratified, effectively nullifying the legislative solution to the strike.[33] With their hopes dashed by the Court, the mill hands went back to work.

That summer, the Orizaba Federación Sindicalista led the workers from the mills' hydroelectric plants out on strike.[34] The State Labor Board had mandated a pay raise that the companies refused to pay.

The Orizaba owners then tried to impose individual labor contracts on their workers, nullifying collective contracts and unions. The CROM led a forty-nine-day walkout, which Obregón and Calles supported. President Carranza left the final decision to the Junta Central de Conciliación y Arbitraje, which ruled in favor of the CROM unions.[35] Defeated by an alliance of a national labor confederation, the CROM, and the state, this was the last time the owners mounted an all-out assault on unionization.

These 1917 and 1918 strike movements suggest the direction of change in trade unions. Regional and national alliances had taken the lead in labor affairs. The overall success of their strikes demonstrated that regional alliances could accomplish things that factory unions could not, and that the new institutional environment favored regional groupings allied with the state or at least with powerful figures in the state. Before 1917, the action in labor affairs had been in factory unions. Organizing the shop floor concentrated workers' activities. After 1917, and especially after 1920, factory unions ceded to regional and national federations. Furthermore, factory unions and labor federations sought allies in government—local, state, and federal—despite the lack of stability to be found there. In textiles, hardly any individual factory union remained outside the larger federations and their links to government.

The shift from factory unions to national labor federations that collaborated with the state fundamentally altered the social relationships of unions, especially and most important, those between leaders and rank and file. Earlier, when unions were not legal and had no money or influence, only the most courageous accepted the responsibility of leadership. Since national organizations were as precarious as factory unions, most local leaders emerged because of their prestige among the factory's mill hands. After 1917, unions enjoyed legal protections, dues collections, and control over hiring and firing, which is to say union officers became protected officials with incomes and power. At the same time, the political influence of national organizations like the CROM and CGT gave an advantage to leaders with national ties as opposed to prestige among local rank and file. They were the ones who could get things done. Furthermore, common workers knew that in case of conflict with the factory, not only would the union defend them, the union leader would control who served on the labor boards that judged cases of disobedience and other transgressions.[36] Thus the increasing power of unions benefited workers but made them more dependent on leaders. Furthermore, as unions gained control over hiring, more and more workers became beholden to the union official who hired them, especially important in a society of personal relationships like that of revolutionary Mexico. While the new formal labor regime strengthened unions, it led to the increased importance of union hierarchy.

Increasingly powerful factory union leaders naturally sought to expand their influence inside the mills. They wanted "to move from one department to another whenever there is the need"[37] and to leave the mills to attend to union affairs.[38] They eventually won these rights through numerous labor actions. It was part of a process in which the position of union leader became a profession. Although virtually all union leaders emanated from the ranks of the common worker, many aspired to moving up within the union, then the federation, and, they hoped, within the national confederation and even into national politics.[39]

As the importance of union officers grew, particularly that of general secretary, these leaders used their power and influence to dominate local politics. This was particularly notable in the mill towns of Orizaba and Atlixco. In the Orizaba region, textile leaders became municipal presidents in Nogales (1914), Orizaba (1916), Santa Rosa (1917), and Rio Blanco (1918).[40] By the 1930s, local labor leader Martín Torres not only served as Orizaba's municipal president, but also as regional strongman. In Atlixco, the leader of the local labor federation, Benito G. León, came to be more powerful than the municipal president. During the conflict between the CROM and the libres, León called on Morones for help. Morones arranged for President Obregón to send an army batallion to put four hundred libres on a train to Mexico City.[41] In 1924, when *sindicalista* Benito Flores became municipal president of Atlixco, he ordered the arrest of Federico Fantini, the Metepec administrator and ordered him to sweep the Zocalo for an hour, which workers saw as "*muy bonito.*"[42]

During this period, union officials put more emphasis on discipline and loyalty among members. They sometimes punished or even fired workers for violating union rules. In 1920, El León complained that the secretary general of the union, not the mill, fined workers as a means of discipline.[43] The fines, three or four days pay, were much like the fines applied by the mills before the workers' revolution.[44] The union also forced a dissident supervisor to quit, in effect firing him.[45] Administration described the Secretary General J. Cruz Rivera as a worker who never spent time at his own job, constantly walking about the mill discussing union affairs and taking for himself the power to hire and fire. He threatened his enemies with murder. The mill concluded that "it is not convenient to continue working with this bad element."[46]

Owners found, however, that they could not get rid of them. To the contrary, with time, the leaders would become even more powerful. In each region, a powerful cacique emerged who ran the unions and the region. Years afterwards, in Atlixco, Antonio J. Hernández "became the leader, and he had his action group kill anybody who wasn't convenient to him."[47] If wealthy owners could not control such men, what could common workers do?

It is important to note that all factory unions and regional and national federations elected all officers. However, the degree to which a free electoral process drove genuine union democracy remains unanswered until we have more detailed studies of the 1920s. At this point, we know that before 1923 unions pursued workers' goals, union leaders came from rank and file, and that conflict inside unions was less than conflict between competing unions or with companies. We also have numerous cases of rank-and-file workers forming quickly elected, ad hoc union commissions to deal with specific problems, suggesting a form of union democracy during this period. When the mill hands in El León became unhappy with cleaning pay, they "precipitously quit their tasks in order to leave to form a new commission, supported by the other departments."[48] These were almost always rank-and-file efforts.

The new power of leaders emanated from institutional and noninstitutional factors. Institutionally, they served on the new labor boards or appointed the union representatives who did. They participated in the political parties that ran local and national politics. Their positions within the unions were protected by law. Owners had to deal with them, like it or not. Common workers also needed them, which increased their subordination. When Camilo Maure discovered workers sabotaging cloth in his mill in 1928, he could not fire them. Instead, he had to present a case to a Labor Board dominated by union leaders. Although the Labor Board fined the workers, it also levied a fine on Maure, and no jobs were lost.[49]

Cultural factors also played an important role. The patron-client relationship had a long history in Mexico. Powerful men were expected to provide favors, and those who received favors were expected to provide support. Union leaders could get jobs for friends and family, protect workers against firings, make life in the mill more pleasant, and solve problems. If they used violence or illicit monies to accomplish their goals, most workers saw it as legitimate insofar as they pursued workers' goals. Furthermore, it was difficult to find any power in Mexico of the 1920s that did not depend on violence and illicit monies. Workers respected power at the service of labor's goals. The new union leaders fulfilled such expectations and made life better for many if not all workers. In fact, the evidence suggests that they made life better for most workers through most of the decade. That is why rank and file mostly supported the new unions and the new leaders. Until 1923, union leaders, factory unions, and labor federations pursued the goals that workers had sought since 1910.

As mentioned, the new wealth and power of unions made them an inevitable target of competing groups, which gave rise to the three-way conflict between CROM, CGT, and Libres. As Barrolo of El León noted when his workers walked out, it was "because of the differences between the Communist and Syndicalist parties, to which they belong."[50] In a strange way, this civil war among unions signified their final victory. First, owners finally realized they could not undo unionization. Instead, they tried to shape it by substituting docile or white unions for the more radical CROM and CGT. Shaping unions, however, meant accepting unions. Second, the premise of the CGT and CROM war was that unionization had won, but what remained to be decided is who would control it. Finally, the intraunion fight meant that unions had become so wealthy and powerful, they were worth dying for and killing for. Curiously, the civil war among unions in cotton textiles represented a victory for the workers' revolution rather than a defeat.

Thus the transformation of trade unions took place during a period of increased unionization and increased violence. Trade unions came to control the entire industry. Unionization even extended to white-collar workers. White-collar unionization began in Santa Rosa in 1916, and other mills quickly followed. In the beginning, the two groups, blue and white collar, were aware that "antagonism exists between us."[51] Over time that changed to cooperation as they affiliated to the same federation. As might be expected, the unionization of white-collar employees meant a further loss of owner control of the mills.

Textile unions also began to organize workers in other industries, including rural workers.[52] In October 1920, Gustavo Hernández, general secretary of the Centro Sindicalista de Obreros del Ramo Textil in Tlaxcala, wrote the state governor to ask for authorization to organize *campesino* unions in nearby haciendas. He needed help because hacienda administrators would not allow him to speak with workers.[53] A week later, campesinos from the Hacienda de San Diego Apatlahuaya, in Santa Cruz Tlaxcala, informed the governor that they had organized a union. They proceeded to work an eight-hour day.[54] The hacienda administrator tried to increase the workload and warned that he would not employ them for only eight hours unless other haciendas did the same. Since that was not the case, those who would not work as before were fired. The newly unionized rural laborers asked the governor to invoke the new Constitution to help them keep their jobs.[55] Unionization in textiles helped spread a culture of labor rights throughout the country.

Control over the Shop Floor

The new institutionalization of labor affairs, the intensification of labor violence, and the transformation of trade unions between 1917 and 1923, did not affect the principal goal of the workers' revolution: gaining control over the shop floor. Workers and unions continued to fight for a better standard of living and more control over the factory. As an angry factory administrator noted, unions sought "direct influence on control and management of work."[56]

To the unions, controlling the factory meant determining who worked there, so that many struggles during this period were about firing, disciplining, and hiring workers and foremen. After the Constitution, however, and particularly after the writing of state labor codes, many of these struggles must be understood as attempts to shape the informal aspects of processes in which the formal labor regime already favored workers and their unions.

Workers had long sought to get rid of certain hated foremen, and they pursued this goal with even greater effect after 1917. In December 1918, the mill hands in La Carolina demanded the company fire the administrator, blaming him for the low quality of output. The company responded by threatening a lockout, but in January the workers walked out. When the Atlixco municipal president intervened on the side of the union, it demonstrated the effectiveness of the new institutional environment.[57]

In August 1920, the El León union threatened to get rid of supervisor Lauro Cruz if he didn't join the union. The union won when Cruz "decided to quit."[58] In Orizaba, when the workers "agreed to expel an *empleado*, they would stop work, surround the man and force him to leave the mill in midst of shouting, whistling, and general tumult."[59]

Through such methods, unions intimidated many supervisors and white-collar workers.

Meanwhile, unions fought to prevent the mills from firing people. There were fewer conflicts over this issue than one might suspect, however, because Article 123 prohibited firing workers for union activities. Furthermore, the state labor codes strengthened protections against firings. Nonetheless, there were cases. In March 1918, workers attacked Abraham Franco, owner of a small workshop in Toluca. Two local labor leaders, Espridio Márquez and Félix Lazo, led a break-in of 150 workers who physically threatened Franco for having fired some of his workers. Although Franco complained to the governor, the case was sent to the local Junta de Conciliacion y Arbitraje in Toluca rather than criminal court, in effect allowing the attack.[60] A culture of unionism and violence prevailed over law.

In addition to control over firing, the unions also sought to extend their control to hiring. On October 20, 1920, Vicente Díaz, general secretary of the first shift union, and Agustín Díaz, general secretary of the second shift union, asked Anacleto Cortés, La Trinidad administrator, to give a job to Abundio Vázquez. Vázquez was an old union hand who had been requesting a position for six weeks. Cortés denied their petition. Vicente and Agustín believed that the administrator was simply boycotting the union. The matter quickly escalated as Cortés threatened to not pay the workers if the union insisted on supporting Vázquez. The union leaders decided to call a strike in order to get rid of Cortés. The strike/lockout began almost immediately.[61] The company claimed that the union enforced the strike/walkout by beating up mill hands who tried to return to their posts.[62] By November, the strike movement had spread to other mills in Tlaxcala and Puebla. Demands expanded from the job offer for Vázquez to the right to hold meetings and collect union dues inside the factory, changes in work assignments and procedures, and the assurance that no worker be fired for participating in the strike.[63] On November 18, when the owners and the Federación Sindicalista of Tlaxcala signed an agreement to end the strike, it appeared that the workers had won:[64]

1. a commitment to review the performance of Anacleto Cortés
2. the right to hold union meetings in the factory during each shift
3. no firings or reprisals
4. worker-approved changes in work rules and procedures.

When the agreement was not fully implemented, the strike movement spread to other local factories, including La Elena, La Estrella, San Luis, and La Tlaxcalateca.[65] Meanwhile, labor federations in Mexico City and the State of Mexico demanded that the Labor Office use its influence to get rid of Cortés.[66] The union enforced the strike by preventing scabs from entering the factory.[67]

On January 8, 1921, the union finally agreed to go back to work after an agreement to reconsider the jobs of both Cortés and Vázquez. Although its victory was less than spectacular, the union had carried out a grand battle to gain control over the shop floor and watched the strike movement spread to nearby mills.[68] Then the Atlixco unions launched similar actions to gain control of hiring and firing in their mills, with 1921 strikes in La Carolina, La Concepción, and the always conflictive El León. In La Concepción, it was because the mill denied "rights to union leaders to hire and fire operators."[69] In El León, the union demanded the mill fire two supervisors and that it be given "direct influence on command and control of work."[70]

The Puebla State Labor Code of 1921 came out of these battles in the mills, and it gave the unions absolute power in hiring and firing. That the unions got this power meant that the union bosses got it. Even so, there can be no doubt that most common workers preferred that workers rather than owners hire and fire. Local union leaders still came from the factory and the communities of workers and shared their values. Of course there could be great conflict in cases of competing union federations, but other than that, workers saw union control as their victory.

Workers and unions also sought other forms of control over the shop floor, the *"ingerencia directa de mando y manejo del trabajo."*[71] In 1920, a year before the Puebla Labor Code, the El León union fired workers hired by the mill, replaced them with unionists, and collected union dues inside the mill. When mill officials protested these actions to the union boss, the aforementioned J. Cruz Rivera, he replied that "they shouldn't be surprised if they get killed."[72] Thus, management had to temporarily close the second-largest mill in Atlixco because the union controlled the mill and threatened management with murder.

The assault on the authority of factory administrators was more or less constant. In January 1920, Mario Leautaud, the administrator of Santa Teresa, was making the rounds when he discovered a weaver, Inocencio Chávez, away from his post. In his report, Leautaud claims to have told Chávez to return to his loom. A half hour later he found the weaver in the factory's cleaning room, so he again ordered the not-so-innocent mill hand to get back to work. Inocencio responded with a sarcastic "he didn't feel like doing what he was ordered." Leautaud then threatened to throw him out of the building. Angered, Chávez pulled a knife and tried to stab Leautaud, who dodged the blade while delivering a blow to the rebellious worker. The municipal delegate, who happened to be in the factory at that moment, helped remove Chávez from the building. They failed to calm him, however, as Chávez continued to scream insults at the French administrator, challenging his manhood and demanding that he step outside to settle the affair.[73]

Leautaud claimed that union officials were responsible for agitating the workers after Chávez's expulsion. A hundred operarios declared a strike at 10 A.M. that morning. Local authorities arrested Chávez and jailed him, prompting the weavers to continue their strike for three days before returning to work, which union leaders Cecilio Nápoles and Carlos González opposed. They wanted to continue the strike until Chávez got his job back. Nápoles was so angered by what he claimed was the weavers' lack of solidarity that he immediately resigned from the union and his job at the mill. González returned to work with the others but resigned his union post.[74] Although the union did not win in this case, it was a notable event for three reasons. First, a three-day work stoppage to defend a comrade who was standing idly in the factory, who refused to obey orders, who tried to murder a supervisor, and who threatened the Frenchman's manhood, was symbolic of the attitudes of workers after a decade of revolution. Second, the mill could not fire any of the protesting mill hands. Third, and significantly, the resignations by Nápoles and González suggest that in 1920 the divorce between union leaders and rank and file was not as large as it would become in later years. In this case, leaders led, or tried to lead, but did not command.

The spirit of revolution exhibited by the Santa Teresa workers traveled quickly from factory to factory. Militant socialists, angry anarchists, confirmed unionists, determined organizers, and disgruntled workers learned about activities in one factory and carried the news to the next. If one worker could get away with refusing an order, so could others. If one mill hand could stab a boss, so could the rest. Radicals and militants and just plain gossips spread the news from region to region. Although factory owners labeled them "professional agitators," they were the voice of revolution inside the factories.

Later in 1920, Luis Sosa, a weaver in La Constancia, came to work drunk, loudly demanding his weekly pay. When the employers refused to pay him, the infuriated mill hand began screaming epithets. The factory administrator tried to remove him from the building but Sosa pulled a knife, the administrator grabbed a pistol, and in the ensuing scuffle, the administrator clubbed Sosa on the head and dragged him from the building. He then returned to the mill to lecture the other workers that "people of this sort couldn't work there," firing Sosa on the spot.[75]

The following morning union leaders were rebuffed when they asked management to readmit Sosa. The workers then walked out, demonstrating that Sosa's alcohol, knife, and threatening words were perfectly acceptable to them, just as Inocencio Chávez's behavior had been acceptable to his comrades.[76] Furthermore, the defense of such behavior was important enough to convince many to lose work time and pay over the issue. This indeed was revolution! The La Constancia walkout indicated the fundamental transition that a decade of violence brought to the cotton textile industry. Workers had created powerful unions, which they used to carry out strikes in defense of knives, alcohol, fisticuffs, and the challenge to authority. They were also willing to strike without warning, to unite with other workers, and to fight to control the shop floor.

At that time, the 800 workers at La Trinidad in Tlaxcala declared a strike because they did not agree with the way management determined piecework rates. The company complained to the Labor Office that whenever it had instituted changes in the past, the workers continued at their tasks while management drew up a new piecework schedule. This time, however, mill hands not only refused to continue working, they neither advised management of the walkout nor provided the company with an explanation; they simply walked out. [77] Following the action, both the company and the union appealed to the Labor Office. With the support of the governor, the Labor Office appointed Antonio Juncos, a member of the Tlaxcala State House of Representatives, to arbitrate the dispute. Juncos found in favor of the workers.[78] Once again, politics, power, and the new institutions favored workers.

In Mirafuentes, the mill hands also protested a change in cloth by walking out. Four days later they returned to their jobs after the mill agreed to a 30 percent wage increase. [79] One hundred three mill hands won higher wages because of their newly powerful unions.

In January 1922, the first-shift workers in La Paz, a medium-size cotton mill in the city of Puebla, entered work at 7:30 A.M. "with their hats on." Pedro Segarra, the administrator, ordered them to remove their hats. The mill hands not only refused, they then began to place pieces of cotton on their heads, arguing that they were cold and would not take the hats or cotton off their heads until 11 o'clock that morning. When Segarra threatened to shut down the plant, "the workers, carrying arms of various types, moved towards the motors and started them, telling the administrator that if he stopped things, 'blood would flow through the factory.'" Miguel Abed, the owner, then ordered the mill closed "until a final arrangement was reached with the workers." After the walkout/lockout, the workers agreed to go back to work without their hats as long as nobody got fired, which Abed accepted. This was another local action for shop-floor control.[80]

Thus, the workers' revolution in 1923 continued what it had begun in 1910, challenging the authority of capital. In 1919, the general secretary of the El León union explained the union's demands that led to a walkout: increased wages, freedom of movement for union officials within the mill, changes in measurements, and a school for workers.[81] The laborite goals were familiar but the methods had changed as guns replaced knives, workers refused to obey orders, and mill hands walked out at the slightest grievance. Rabid owners came to admit that they had lost control of the shop floor. If workers set out to destroy the authority of bosses in 1910, by 1923 they had won. The workers' revolution was violent and unclean but not unsuccessful.

CHAPTER 9

The Revolution and the Labor Regime, 1910–1923

MANY YEARS BEFORE I BEGAN this manuscript, I taught a class at the Escuela Nacional de Antropología e Historia in Mexico City with Adolfo Gilly. I was very young at the time and had not yet completed my Ph.D., while Adolfo had recently left the prison where he wrote his landmark study, *La Revolución Interrumpida*. Among the very bright students in that group, one asked whether Mexico's revolution measured up to the classic cases of France and Russia. Adolfo provided a response that I never forgot. A revolution, he said, is not what the classic textbooks say a revolution should be, it is what the people do. *"La revolución no es lo que dicen los textos, sino lo que hace la gente."*

Years later when I started this project, scholars had begun to question whether Mexico's revolution was even a social revolution. They followed a trend that questioned social revolution in France and Russia, the classic cases. While working in the archives for this project, such questions made little sense. During the revolution itself, many people discussed its value but no one doubted that there had been one. In 1919, F. Harvey Middleton wrote:

The overthrowing of the old regime has, it is true, resulted in much unwise legislation, in some cases jeopardizing the industries created by American and European capital and energy—industries upon which Mexico is absolutely dependent. But the old order of things in Mexico, with single families owning millions of acres and a dozen families owning entire states, has gone never to return. You can no more restore the Mexico of Porfirio Díaz than you can bring back the Russia of the Czars or the France of the days of Mme. de Pompadour.[1]

Indeed, even a comparison of the photographs of the well-groomed lawyers and businessmen who ran Mexico before the revolution with those of the rough-hewn and violent men who ran it afterwards, found in Anita Brenner's classic photo history, suggest the depth of change in the country.[2] Mexico may not have been France or Russia, but it was revolutionary.

Prevailing opinion after the revolution argued one of three positions. First, it was a revolution about land, symbolized by the iconic figures of Emiliano Zapata and Francisco Villa. Second, it was a revolution among competing elites, represented by the no less iconic figures of Francisco Madero, Venustiano Carranza, and Alvaro Obregón. In more recent times, some scholars have argued that rather than a national revolution, there were different regional revolutions with their own causes and processes.[3] Without doubt, there is some truth to each of these.

Within that debate, nobody argued that the Mexican revolution was a workers' revolution, although the industrial working class in Mexico was no less important in its society than the industrial working class in revolutionary Russia. An exception was that near contemporary, Ernest Gruening, who wrote, "the most palpable product of the Mexican Revolution is the labor movement."[4] Gruening, Marjorie Ruth Clark, Vicente Lombardo Toledano, and others had pointed to labor's outcomes: Article 123, strong unions, status within the ruling party. But everybody insisted that labor's gains came from a benevolent state, which hardly seems plausible in the rough-and-tumble world of postrevolutionary Mexico. How could workers have won so much if they had done nothing while the country around them burned? Were the generals and politicians who took power in the late revolution really that kindly? These questions led me down a research path that uncovered a workers' revolution within the Mexican revolution.

A social revolution at work is a fundamental change in the labor regime. The labor regime is the set of formal and informal institutions that define the social relationships of work, and as a derivative, the social division of the products of work. The formal components are the written laws, work rules, labor contracts, and organizations like trade unions, employer associations, and government offices. Equally important are informal institutions such as the embedded rules, habits, and norms of work and workplace relationships. Among these, none is more important than authority at work. Authority is the central question of every social revolution. When those at the bottom do not question those at the top, there is no revolution. When those at the bottom no longer accept the authority of those above them, when they think they have the real possibility of challenging that authority, and when they act on that belief, by definition there is a social revolution. The revolution ends when a new authority at work imposes itself. That is the reason why the labor regime, which focuses on authority at work, is a better category for understanding the workers' revolution than moral economy, a broader term that encompasses other social processes.

For there to have been a social revolution among workers, there would have to have been a revolutionary challenge to the traditional authority at work. I have tried to demonstrate that there was, not only in my view, but more important, in that of workers and owners. Operarios, with firm ideas about social class, challenged authority at work from the beginning of their revolution until the end. However, the nature of their challenge differed from that of their Russian counterparts, which is partly why it was overlooked. In Russia, anarchist, socialist, and Marxist political parties questioned the rights of capital. Numerous workers participated in those parties and acted on anarchist, Marxist, and socialist ideas. As Chapter 5 argues, such ideas and political parties were mostly absent from Mexico's textile zones. Instead, Mexican workers engaged in a more empirical revolution, with one set of events leading to another. The breakdown of Porfirian political rule al- lowed workers to successfully challenge owners on traditional laborite issues like wages and hours of work, the subject of Chapter 4.[5] A lack of effective repression then allowed workers to successfully challenge owners on other issues, particularly authority at work, as seen in Chapters 5 and 8. It was an empirical process, each battle leading workers to draw their own conclusions, each victory strengthening their resolve, each defeat not strong enough to stop their growing ambition to control as much of the shop floor as they could.

This challenge to authority at work became the central issue of Mexico's workers' revolution because it transformed a laborite struggle into a revolutionary struggle for control of the factory. Before the revolution, workers won some battles but mostly they lost. They lacked strength and organization, their methods and goals were often timid, and the owners could usually count on local and national authorities to repress them. Typical was the conflict in 1889 at La Carolina where the mill won because it was able to fire the rebellious workers, replace them with strikebreakers, and count on the government and the armed forces to maintain order.[6] Backed by a strong state, administrators controlled who worked in the mills. The law, the rights of private property, and the strength of owners relative to the weakness of workers determined factory control over hiring, firing, and the work process. As in most places, the old labor regime rested on respect for the authority of capital.

The 1911 general strike was the critical turning point in Mexican labor history. It was a widespread and unprecedented victory for workers that began as a laborite struggle but also as a challenge to authority. Although workers agreed on wage and work day demands, in many factories mill hands added local petitions about dignity and workplace control. In Atlixco, it was the demand to control one's house, at La Carolina, strip searches, at La Hormiga, the size of cloth. The strike's leadership was careful to argue that workers' demands were "strictly just and . . . within the limits of the rational."[7]

Winning the strike increased self-confidence, which in turn led to more strikes and workplace demands, as when the weavers of El Carmen walked out in order to get an assistant to carry the cloth rather than carry it themselves.[8] After 1912, such local actions multiplied, leading to the strike wave of late 1912 and early 1913. Workers walked out again and again, the new Labor Office simply noting that "the cotton textile workers have expressed their dissatisfaction and unhappiness."[9] This phase of revolution strengthened workers' belief in the trade unions that had brought them success. Speaking for all textile workers, San Antonio Abad mill hands declared that they had been "victims of our owners and bosses,"[10] capturing their sense of class and their political program: take power away from their victimizers. Of course, a broader revolution that had weakened government enabled these victories, as workers were aware.

The collapse of central government following the assassination of Madero, the spreading civil war, and the anarchy of a country governed more by regional military commanders than by centralized civil authority, further emboldened workers and unions. Elites, engaged in their own fight for survival, were simply unable to mount an effective repression of an increasingly rebellious working class. Between 1913 and 1917, workers attacked supervisors and walked out or engaged in other work actions to get what they wanted. Factories found it impossible to repress the labor actions. The administration of Cocolapam characterized the protests as "the infantile and impulsive movements that have spread disorder in the factory."[11] The workers naturally had a different opinion, telling Antonio Valero, head of the Labor Office, "And let the Industrialists know once and for all that the epoch of tyrannies has ended."[12] Santa Rosa workers understood that "the desires of the Triumphant Revolution of the People" had vindicated "the pariahs and [returned] their rights to them."[13]

As central government collapsed, regional military commanders responded to the workers' challenge to authority with military decrees that ceded many of their demands. The supreme decree, of course, was Article 123 of the 1917 Constitution. Protected unions, legalized strikes, collective contracts, and numerous benefits made the revolutionary Constitution the legal victory of the workers' revolution. Interestingly, once workers began down the road of greater workplace power, each victory seemed only to whet their appetite, so that the Constitution and the subsequent state labor codes increased rather than diminished the challenge to authority, as seen in Chapter 8.

After the strike at San Juan Amatlán, the mill owners, astute businessmen if not scholars, said that hegemony in the workplace rested on "the principle of authority," which in turn underlay the hegemony of the state. If workers could challenge authority at work, as happened at Amatlán, then it would destroy the "the only possible basis on which the State can sustain itself."[14] Without the benefit of Gramsci, they understood their need for a state that demanded respect for private property, the unquestioned authority of capital inside the factory. That, however, was not the Mexican state in the 1920s.

Thus the pervasive, persistent, and often violent challenge to factory authority led industrialists in 1920 to beg *el Supremo Gobierno* to "establish an industrial peace in the Republic, removing or making less frequent, the strikes."[15] They were unsuccessful. The uncontrolled labor actions of the workers' revolution forced a new state to adapt to workers as much as workers adapted to it. There were sufficient activist mill hands carrying out what the owners labeled the "the biased and perverse goals of the agitators" to stalemate the industrialists and a weak government, forcing the construction of a labor relations systems that favored unions and a labor regime that favored workers.[16] Uncontrollable shop floor conflict changed the labor regime in Mexico because, as one owner noted, "It is very difficult to govern the factories and technically organize production when these anarchic conditions prevail and the orders of the administrators and the supervisors are ignored."[17] The revolutionaries who took power in Mexico changed labor institutions because they had to. Mexican factory workers refused to obey and work as before. That is why the labor outcomes of the 1920s came from the workers' revolution rather than from a benevolent state.

The new labor regime and its real and fundamental improvements in the lives of workers created a new hegemony of the social relations of work in postrevolutionary Mexico. The new hegemony, whose story is better left to another volume, is what ended the workers' revolution. There was no decisive defeat by capital, at least not by 1923.

This is not to argue that cotton textile workers were a homogeneous group who always fought authority at all times and all places, because that was certainly not the case. However, that was not necessary in order for workers to win. Given the context of the surrounding social revolution and the weakness of central government, a sufficient number of workers fought authority in enough instances to completely transform the social relations of work in Mexico.

If many revolutionaries and scholars later argued that even in Bolshevik Russia a workers' state never emerged, in Mexico it certainly did not. A proworker labor regime meant that workers won, but there were limits to their victory in three critical areas. Before looking at these limits, however, let us review some of the major victories of the new labor regime, at least to the mid 1920s.

Workers started their revolution over wages and hours of work. In Mexico, it is difficult for real wages to rise for long periods because of the pressures of an underdeveloped labor market. From the late Porfiriato to the early twenty-first century, rather than a steady increase in real wages based on the rising productivity of labor, there are cyclical periods of rising wages followed by sharply declining wages, with little net overall increase. During the late Porfiriato and the early years of revolution, real wages in textiles fell. From the late revolution to the 1930s, real wages and benefits increased, clearly as a consequence of the new labor regime.[18] Aurora Gomez suggests that real wages in cotton textiles rose sevenfold between 1916 and 1928.[19] Stephen Haber calculated a 43 percent increase between 1925 and 1929.[20] In fact, the only periods of sustained increase in real wages in Mexican industry were from the late teens to the late 1930s, a period during which the institutional factors of the workers' revolution influenced real wage growth, and from 1952 to 1975, which coincided with the general ascent of growth rates in the world market, which pulled Mexican industry upwards. The workers' revolution increased incomes to textile workers for an important period.

Mill hands also won significant reductions in the workday. In December 1906, the recently established Centro Industrial Mexicano unilaterally instituted a fourteen-hour workday. Workers could leave at six on Saturdays, when they enjoyed a reduced workday, 12 hours.[21] Mill hands protested, struck—leading to the famous events at Río Blanco—and lost. The power of owners prevailed, so that the prevailing workday at the outbreak of revolution in 1910 was fourteen hours.[22] The 1911 general strike reduced the workday to ten hours for the day shift and nine for the night shift. Between the January 1912 provisional settlement and the July contract, there were strikes in a number of factories to make them comply with the ten-hour workday.[23] Following the July contract, the strike wave grew as mill hands insisted on the enforcement of the contract's provisions. Ultimately, mills fell into line with regard to the new workday. Federal labor inspectors fanned out over the country, confirming the change in working conditions.[24]

Conflict over the workday was not limited to the textile industry. During a 1913 railroad strike, workers not only expressed pride in shutting down the trains, but claimed that if successful, they would demand a workday of six hours within two or three years.[25] Many of the military decrees from 1914 to 1916 reduced the workday. In September 1914, Pablo González, issued a decree for Puebla establishing the eight-hour day.[26] In October 1914, Gertrudis G. Sánchez issued a decree setting the maximum workday at nine hours.[27] At a union meeting in Orizaba, one mill worker reminded his comrades of the importance of "resting, cleaning up, and amusing oneself."[28] By 1915, the Labor Office noted that "the nine and eight hour days . . . are well accepted by the workers."[29] Article 123 ratified the eight-hour day. Thus the workers' revolution increased wages and benefits while cutting the work day almost in half.

Health care is an example of improved benefits. Even if life in the factory was not necessarily a worse alternative than life in the fields, textile work caused health problems. A 1919 inspection of La Unión reported:

All of the machinery of the factory is installed in one large room that measures about sixty meters in length by twenty-five meters in width . . . With respect to the machines that are working they are in equally deficient condition because the transmission bands are very old and patched which causes the machines to not work as they should, being also in my opinion a constant danger to the workers. The machines called *pulidoras* constantly create a strong current of air when they are working that goes directly to the worker; the manager himself told me that almost all the workers that operate these machines are constantly getting sick from *pulmonia* and some have even died from the disease.[30]

In the much larger Río Blanco factory in Orizaba, management listed bronchitis as the major medical problem.[31] A complaint in La Purísima noted that the carders were too close together, making it easy to cut off a finger or a hand.[32]

During the Porfiriato, there were no legal requirements for factories to pay medical expenses, though some did. The 1906 CIM rules ignored health issues. Article 18 of the 1912 contract, however, made the owners responsible for free medical care in case of work accidents. Following the strikes of 1911–13, workers protested insufficient medical attention, leading to military decrees that often mandated medical coverage. Cándido Aguilar's Decreto Número 11, Article 7, required that owners provide medical care for workers and to pay their salaries while out sick.[33] When the Villistas and Zapatistas controlled Mexico City, they agreed that workers' health was a major concern, signing an agreement that included medical provisions "that makes less cruel the exploitation of the proletariat."[34] Article 123, Chapter XIV, made owners responsible for medical care for job-related illnesses and accidents. In the textile states, subsequent state labor codes strengthened medical provisions. In Puebla, there were thirty articles covering medical care.[35] The workers' revolution dramatically strengthened medical care for workers and their families.

Of course, the workers' revolution went far beyond traditional labor issues like hours and wages of work and medical care. The workers' revolution evolved into a struggle for the shop floor, which for unions meant hiring, firing, and disciplining. During the nineteenth century, factory blacklists kept undesirables from the mills. The 1911 strike targeted them, so that at the 1912 convention, Rafael Hernández, Madero's minister of development, noted:

There is, nonetheless, among the petitions of the workers one that I would especially like to call to your attention.

Some delegates from various factories have approached the Minister of Government and me, declaring that there exist a large number of workers who are systematically excluded from work, without there being any reason that justifies this conduct.

They attribute this exclusion to their name figuring on a list, a kind of "Index" that the industrialists have and communicate among themselves, of all those workers that form part of the nascent workers societies and who take a more active and direct participation in them than the rest of their compañeros.[36]

Workers' activism convinced government officials to end their support of the blacklists, and by early 1913 the Labor Department noted the spontaneous and successful organization of unions in the mills. Factory owners, however, tried to maintain "the old system in the factories to fire Mexican workers without motive or any reason." In 1915, Labor Department inspectors found blacklists against workers who "demand their rights, ask for justice, and form part of the union leadership." The factories, they said, "make lists with their names, circulate them among factory administrators so that they will not find work in any factory."[37] Unions fought the blacklists and at-will firings. A regular demand of every strike and walkout was that nobody lose his or her job for walking out. During the May 1916 general strike which paralyzed Mexico City, the four-point agreement to end it included "none of the workers who participated in this strike will be fired."[38]

During the early revolution, labor activists concentrated on curtailing the owners' right to fire. During the later revolution, they tried to influence hiring. While the 1917 Constitution protected workers against firings, it left hiring to the owners. Under union pressure, state legislatures moved further with the state codes. Although the Veracruz labor code allowed owners to hire, following regional strikes between 1918 and 1920, the 1921 Puebla state code gave that power to unions.[39] It also made firing workers very difficult because the new state law recognized bilateral and tripartite labor councils in factories and in the state government, with numerous rights to appeal. An owner could not simply fire a worker but rather had to work through the unions and numerous legal requirements. Going further, the 1927 textile contract, which replaced the 1912 agreements, stated:

Any worker who wishes to be admitted to a factory must make the request by himself or through the representative of the factory's Union. The admission request will be resolved by mutual agreement between the owner or his representative and the respective Union delegate . . . the worker will present to the factory administration a supporting document that he has affiliated with the corresponding Union, not being able to be definitively admitted to the job, although he has fulfilled the first two prerequisites, if he doesn't present this supporting document.[40]

From 1917 to 1923, unions increasingly fought to control hiring, and sometimes they won. With the second national textile contract in 1927, unions achieved the exclusive right to hire. The contract also strengthened workers' protections against firings. It contained an entire chapter, IV, on sanctions. Article 78 established that the first time a worker committed a contract violation, the factory had to ask the union to discipline him. The second time a violation was committed, Article 79 allowed the factory to suspend the worker for a week, while Article 80 allowed the factory to fire workers for a third violation within 120 days or for other grievous acts. In other words, a worker could commit violations like stealing cloth or punching a supervisor twice every four months. However, Article 81 did not allow the sanctions of Article 80 to take effect without the participation of the Comisión Mixta de Distrito, comprising three representatives of the owners and three of the unions. [41] If the Comisión Mixta de Distrito could not resolve the problem to the satisfaction of both sides, there remained the Comisión Mixta Nacional, which also had three labor and three owner representatives. If the Comisión Mixta Nacional could not agree on a solution, then each side could go to court.[42] If a union chose to defend a worker accused of violating the contract or disobeying an order, the procedure to fire the worker was long, complex, and costly for the company, without any guarantee of success. In short, the revolution made it legally very, very difficult to discipline a worker without consent of the union. In fact, it was virtually impossible.

There was an interesting case in July 1927, which takes us a bit beyond the boundaries of this volume but shows the extent to which unions gained control over hiring, firing, and disciplining. That month the administration of Metepec tried to fire seven employees, all of whom were deficient in their work. One of them, Francisco Gómez Moncalian, was secretary general of the Employees Union. The company had little nice to say about him. At age 52, he was described by the administration as "of middling knowledge, one might say inept . . . incapable of carrying out work that requires great care or dedication." Administration complained that he was always reading "magazines and . . . other printed material" instead of carrying out his tasks.[43]

When the company became aware of widespread stealing in the factory, it found that Gómez Moncalian and the other unionized employees who were supposed to control the workers, collaborated with them instead. This would not have happened before the revolution, with its great divide between white-collar workers like Gómez Moncalian and blue-collar operatives. By the early 1920s, however, unionization of both and a new era of proletarians against capitalists turned former antagonists into allies. Because of this, the company believed that it had lost control of the mill, arguing that, "we now believe that we need to replace all our personnel, including some supervisors."[44]

The company tried to fire Gómez Moncalian and Juan Bilbao Bengoechea, but they simply refused to leave their jobs. Metepec discovered that it could not remove them from work, even for theft. The explanation, according to the company, was that it was impossible to remove an employee if the workers' union supported him.[45] The company further argued that the workers' union backed the rebel employees in order to have them "subordinate to the workers unions . . . making good administration of this company impossible."[46] A year later, Gomez's case was still the subject of litigation in the Junta Federal de Conciliación y Arbitraje. Meanwhile, Gómez retained his job, did not go to jail for defying authority, and participated actively in union activities. Incidents such as this convinced the administrators that they no longer controlled the mills.

Before the revolution, owners ran the factory. They hired who they wanted and they fired any worker who did not suit their needs. Through their revolution, workers challenged the owners' right to run the factory. With work actions ranging from strikes to murder, they took that right away from the owners. By the 1920s, management had lost control of the mills, which management itself recognized. Unions hired and, in practice, unions determined who was fired. This was a fundamental revolution in Mexico's labor regime.

Despite workers' great victory, there were three critical areas that established limits to what they could win. The first was the change in the relationship between union officials and rank-and-file workers. Gregory Crider has argued that the change from owners to unions in hiring and firing was not necessarily revolutionary.[47] In the Atlixco mills, "Individualized relations of dominance and obligation were reproduced at every level of hiring category and shop floor hierarchy; this series of personalistic relations defined the sindicato patronage system that emerged in Atlixco's textile mills and communities during the 1920s and 1930s."[48] It was a new politics of "domination and submission."[49] The paternalism of union leaders replaced the paternalism of owners.

While Crider's observations have merit, it is less clear that this negates the value of what workers won. For rank and file, it made a difference that workers serving as union officials controlled a patronage system instead of owners. After all, such men shared the layered communities through which workers defined their lives. The new system provided mill hands with higher wages, greater benefits, increased job security, freedom from harassment by supervisors, and a more beneficial rhythm of work. To get and keep a job, one needed to appeal to another worker instead of the old bosses. The benefits of this system in the 1920s and 1930s demonstrated that union-controlled patronage was better for workers than owner-controlled patronage. Furthermore, unions by themselves could not change popular culture, particularly a very, very old idea of the patrón (powerful person)-client (weaker person) relationship. Unions reproduced inside what they saw outside, hierarchy and control, but that does not diminish the change in the labor regime in Mexico's textile mills. Owners lost power; unions gained power.

In the mid-1920s, Gruening wrote about the Orizaba mills:

In these last mills labor is ruinously undisciplined. The Convencion Textil has not served because labor does not do its part. The abuse is worst in the four mills of the French-owned Compania Industrial de Orizaba, employing 5000 workers, one of which at Rio Blanco was the shambles of 1907. From the ill treatment of the workers twenty years ago the pendulum has swung far the other way. The men often refuse to do as they are told, and it is not possible to discharge a man without the union's consent. Not infrequently weavers, when the spirit moves them, quit their machines for the street, to sip a locally brewed cervecita, leaving their looms unattended. If while they are away a thread breaks and no one is there to tie it, the resulting effect spoils the cloth. The fines which formerly would have resulted are now forbidden under the Vera Cruz labor law. The management has no redress. In 1926 the company's steadily dwindling dividend was cut from six to four per cent. It faces further reduction unless its labor begins sensing its responsibilities.[50]

Thus, there is no question whether capital lost control of the mills. It did. The question is whether, in the process, workers lost control of their unions.

At this point we run into a problem in the historiography. Most modern Mexican labor historians agree that the twentieth-century Mexican political system was corporatist or at least state controlled.[51] The government dominated the powerful labor centrals, which dominated the factory unions, which in turn dominated workers, because if owners couldn't fire workers, unions could. While this paradigm seems a fairly accurate description for the 1950s or 1960s, does it apply to the 1920s? Does Crider's argument apply to an earlier period?

The evidence to 1923 suggests that workers remained mostly in control of their factory unions in cotton textiles, despite the changing relationships between union officers and common workers. Without doubt, union democracy was never perfect in the conditions of revolutionary Mexico. During this period, one can begin to see some separation of leaders from rank and file, particularly in the regional and national federations. How this evolved after 1923 will require further research. Nonetheless, it is equally clear that workers could satisfy their demands through their factory unions, that these unions easily mobilized workers on short notice to defend comrades, and that union leaders all came from the factory and the community. These facts explain the overall absence of challenges to the legitimacy of union officials during the early institutional period, even as leaders acquired greater power and income. Nonetheless, the difference between leaders and followers in 1923 was much greater than in 1910.

The second area was law. The main purpose of trade unions is to sign and enforce collective contracts. Collective contracts are a subset of contract law, which in turn is a subset of the state. Workers could not win a revolution through trade unions without winning the right to have those organizations be party to collective contracts. The 1912 accords began a process of collective labor agreements that evolved through the 1917 Constitution, and then the state labor codes, and solidified with a new contract in 1927. These laws and contracts immeasurably fortified unions, which in turn provided workers with better working and living conditions. On the other hand, they equally fortified the state from which they emerged.

In order to repress workers before the revolution, it was sufficient for recalcitrant owners to count on a strong state. For workers, that meant long hours, low pay, minimal benefits, at-will firings, and mistreatment by foremen and supervisors. What changed this was the collapse of the state during Mexico's revolution, which in turn permitted workers' activism to create and sustain successful and powerful unions. Ultimately, however, these unions needed a strong state to ensure their own strength and survival. What this meant in practice was that unions needed state and federal law to determine their legality and existence. They won this between 1917 and 1923 as an emerging state favored workers and unions. The state and the unions grew together, each supporting the other. There could be no guarantee, however, that a newly strengthened state would not once again ally with capital to crush workers, as had been the case during the Porfiriato. That, however, is a story that would take us well beyond 1923.

The third area is the inherent power of capital. Throughout the revolution, workers fought against owners and supervisors until they won. Even alleged incompetents and thieves could not be removed from the factory. With owners unable to choose their labor force, unable to fire bad workers, unable to even select foremen, it is clear that

Short commings of unions

capital lost control of the shop floor, indeed, of the factory and to a certain degree of the production process. On the other hand, there is no evidence whatsoever that the vast majority of workers intended to supplant capital. There were no demands to turn the factories over to the workers or to have a revolutionary state take ownership. When Carranza backed off the factory seizures, workers fought for shop-floor control but not ownership. The workers' revolution within the revolution was curiously anticapitalist without being socialist or truly anarchist. Thus, while many mill hands spoke and acted with force, even violence, against owners and supervisors, equally common were the voices of those who referred to themselves as "the humble but sincere workers of the factory."[52]

As long as capital owned the mills, workers needed owners as much as owners needed workers. Even one of the extraordinary labor leaders in Orizaba asked the Labor Office in 1913 "to protect the rich patrimony that is the job that I have in this factory."[53] Similarly, "the Humble Servants of the La Guia Factory" wrote that "we are not acting in bad faith and we don't desire bad to anybody."[54] The humble servants of the country's mills turned their world upside down but as a class could never conceive of a society in which servant ruled master. There was no Lenin or Trotsky or Bolshevik Party here, only angry and radical workers with their trade unions trying to control the shop floor but leaving capital to capital. That was the limit of their challenge to authority and therefore the limit to their revolution. It may seem strange to those who study the Russian, Chinese, or Cuban revolutions, but that is how Mexican textile workers conceived of their world.

Given these limits, what difference did a revolution make? A radical difference. Cotton textile workers thoroughly transformed the Mexican labor regime. In 1923, the social expectations of the labor process of both owners and of workers were radically different than in 1910. Owners claimed that they had lost control of their factories, a claim confirmed by independent witnesses. Workers demanded and saw the state enforce rights that made their working lives less onerous than before. They won strong unions, the right to strike, control over hiring and firing, influence over the rhythm of work, the ability to get jobs for friends and family, better medical care and educational opportunities, and new social respect. For a long period, wages and benefits grew, hours of work shrank, and workers and their families enjoyed the improved standard of living and better working conditions. Unions and labor leaders wielded great power. It was not a perfect system, but it did create labor peace.

If hegemony includes, as Antonio Gramsci argued, the "spontaneous consent given by the great masses . . . to the general direction imposed on social life," it is almost certainly the case that the new labor regime became a fundamental building bloc in postrevolutionary hegemony in Mexico.[55] It is one of the reasons that Mexico did not experience the cycles of left-wing governments, military coups, and guerrilla warfare that dominated Latin America in the 1950s, 1960s, 1970s, and 1980s.[56] Life had improved for the country's industrial workers, and they knew that. It was a labor regime whose complexities and contradictions forced owners, labor leaders, government officials, and workers to rely on industrial protectionism to pay for the system and often corruption to make it work. It worked well until globalization killed it at the end of the century, but while alive, its strengths were its limits and its limits were its strengths.

REFERENCE MATTER

Abbreviations, Archives, and Newspapers

Some of the archives I consulted are public and quite well organized, like the Archivo General de la Nación. Similarly, the state archives in Puebla and Veracruz are well organized, as are the collections in the larger municipios like Puebla and Orizaba. In all cases, the people working in the archives were very kind and helpful. Atlixco is quite well organized thanks to the work of Mariano Torres, as is the Santa Rosa union archive, thanks to Bernardo Garcia. In the other municipios, as well as the collections of current and former labor organizations, one is really consulting papers, files of papers, or boxes of papers, and in the case of the old Rio Blanco union, mounds of paper scattered on the floor, so that referencing is trickier.

Many of the revolutionary newspapers are held by the U.S. Library of Congress, where Barbara Tenenbaum has helped so many scholars through the years.

ARCHIVES AND ABBREVIATIONS

Abad de Santillan, International Institute of Social History, Amsterdam
AGEV—Archivo General del Estado de Veracruz
    Ramos:
        DEPS—Departamento de Economía y Previsión Social
        JCCA—Junta Central de Conciliación y Arbitraje
        TPS—Trabajo y Previsión Social
AGMP—Archivo General del Municipio de Puebla
AGN—Archivo General de la Nación, Mexico
    Ramos:
        Archivo Gonzalo N. Robles
        DT—Departamento del Trabajo
        Gobernación

AHEM—Archivo Histórico del Estado de México
AMA—Archivo Municipal de Atlixco
AMCM—Archivo Muncipal de Ciudad Mendoza
AMO—Archivo Municipal de Orizaba
AMRB—Archivo Municipal de Río Blanco
Archivo del Catedral Sn Miguel Arcangel, Orizaba
Archivo CROM Atlixco
    SG—Secretaría General
ASRB—Archivo Sindical de Río Blanco
ASSR—Archivo Sindical de Santa Rosa
CGT—Confederación General de Trabajadores
CROM—Confederación Regional Obrera Mexicana
Exp.—Expediente
INEGI—Instituto Nacional de Estadística, Geografía e Informática
SICT—Secretaría de Industria, Comercio y Trabajo
STyPS—Secretaría del Trabajo y Previsión Social

NEWSPAPERS

La Convención
El Demócrata
El Dictamen
El Imparcial
El Sindicalista
El Universal

Notes

CHAPTER ONE

1. To this one might add the guerrilla movements in the Andean countries and the U.S. military interventions in the Caribbean.

2. Ernest Gruening, *Mexico and Its Heritage* (New York, 1928), 335.

3. Ramon Ruiz, *Labor and the Ambivalent Revolutionaries, Mexico, 1911–1923* (Baltimore, 1976), 81.

4. John Lear, *Workers, Neighbors, and Citizens; The Revolution in Mexico City* (Lincoln, 2001), 362.

5. Salvador Novo, *The War of the Fatties and Other Stories from Aztec History* (Austin, 1994), xl.

6. Vicente Lombardo Toledano, *La Libertad Sindical en México* (Mexico, 1926), 16.

7. Alan Knight, *The Mexican Revolution* (Nebraska, 1990), vol. 1, 2.

8. Theda Skocpol, *States and Social Revolutions: A Comparative Analysis of France, Russia and China* (Cambridge, 1979).

9. John Hart, *Anarchism and the Mexican Working Class, 1860–1931* (Austin, 1978).

10. Barry Carr, *Marxism and Communism in Twentieth-Century Mexico* (Lincoln, 1992); and Manuel Reyna Munoz, *La CROM y la CSUM en la Industria Textil (1928–1932)* (Mexico City, 1988).

11. On the early unions, Juan Felipe Leal, *Del Mutualismo al Sindicalismo en México: 1843:1910* (Mexico, 1991); and Jorge Basurto, *El Proletariado Industrial en México (1850–1930)* (Mexico, 1975).

12. An important examination of workers in the Russian Revolution is in Daniel Kaiser, editor, *The Workers Revolution in Russia, 1917* (Cambridge, 1990).

13. William Rosenberg, "Russian Labor and Bolshevik Power," in Kaiser, 1990, 100.

14. Robert McKean, *St Petersburg between the Revolutions* (New Haven, 1990), 253; and S. A. Smith, "Workers and Supervisors: St. Petersburg 1905–1917 and Shanghai 1895–1927," *Past and Present* (Oxford, 1993), 139.

15. Stephen Haber, Armando Razo, and Noel Maurer, *The Politics of Property Rights: Political Instability, Credible Commitments, and Economic Growth in Mexico, 1876–1929* (Cambridge University Press, 2004).

16. The classic study of industrial workers during the Porfiriato is Rodney D. Anderson, *Outcasts in Their Own Land; Mexican Industrial Workers, 1906–1911* (DeKalb, 1976); and Ruiz, 1976, chapters 1 and 2. A more modern view, though it only includes the early Porfiriato, is Mario Trujillo Bolio, *Operarios Fabriles en el Valle de México, 1864–1884* (Mexico, 1997).

17. Although later economists attempted to dispute his central argument, W. Arthur Lewis did provide an accurate description of the challenge to "Economic Development with Unlimited Supplies of Labour." *Manchester School* 22 (May 1954): 139–91. The unfavorable relationships between the supply and demand for labor in the modern sector depressed wages in Mexican industry, just as it did in the rest of Latin America.

18. Kevin J. Middlebrook, *The Paradox of Revolution: Labor, the State, and Authoritarianism in Mexico* (Baltimore, 1995), 77.

19. There is an extensive historiography of Mexico's revolution, but the modern reader would certainly begin with Alan Knight, *The Mexican Revolution* (Lincoln, 1990); Friedrich Katz, *The Life and Times of Pancho Villa* (Stanford, 1998); and Haber, Razo, and Maurer (2004). See also Adolfo Gilly, *La Revolución Interrumpida* (Mexico, 1971, 1994); John Mason Hart, *Revolutionary Mexico* (Berkeley, 1987); and Ramon Ruiz, *The Great Rebellion, Mexico 1905–1924* (New York, 1980).

20. M. Meyer, W. Sherman, and S. Deed, *The Course of Mexican History* (Oxford, 1999), 417.

21. Nevin O. Winter, *Mexico and Her People of To-Day* (Boston, 1907), 202.

22. C. Reginald Enock, *Mexico, Its Ancient and Modern Civilisation, History and Politcal Conditions, Topography and Natural Resources, Industries and General Development* (London, 1909), 167.

23. José Fernández Rojas, *La Revolución Mexicana de Porfirio Díaz a Victoriano Huerta 1910–1913* (Mexico, 1913), 96.

24. Ibid.

25. Katz, 1998, 198.

26. Meyer et al., 1999, 516.

27. John Womack, Jr., "The Mexican Revolution, 1910–1920," in Leslie Bethell, *Mexico Since Independence* (Cambridge, 1991), 154.

28. David Gordon, Richard Edwards, and Michael Reich, *Segmented Work, Divided Workers, the Historical Transformation of Labor in the United States* (Cambridge, 1985), 25.

29. Andrew Sayer, Moral Economy, published online by the Department of Sociology, Lancaster University, http://www.comp.lancs.ac.uk/sociology/papers/sayer-moral, p 1.

30. James C. Scott, *Weapons of the Weak: Everyday Forms of Peasant Resistance* (New Haven: Yale University Press, 1985).

31. Stephen Hill, *Competition and Control at Work* (London, 1982), 259.

32. Gordon et al., 1985, 25.

CHAPTER TWO

1. R. J. Peake, *Cotton* (London: Sir Isaac Pitman & Sons, n.d.), 1.

2. Or alternately, *qutn. The Random House Dictionary of the English Language* (New York, 1983), 331; and John Ayto, *Dictionary of Word Origins* (New York, 1991), 139.

3. Advanced and backward refer only to labor productivity.

4. William B. Husband, *Revolution in the Factory, the Birth of the Soviet Textile Industry, 1917–1920* (Oxford, 1990), 7.

5. Sujata Patel, *The Making of Industrial Relations, The Ahmedabad Textile Industry 1918–1939* (Delhi, 1987), 1.

6. Hernán Cortés, "The First Letter," in Hernán Cortés, *Letters From Mexico* (Yale, 1986), 30.

7. David J. McCreery, *The Sweat of Their Brow; A History of Work in Latin America* (Armonk, 2000), 68.

8. Richard J. Salvucci, *Textiles y Capitalismo en México* (Mexico, 1992), 88.

9. McCreery, 2000, 69.

10. Salvucci, 1992, 29.

11. Jesús Rivero Quijano, *La Revolución Industrial y La Industria Textil en México* (Mexico, 1990), 100.

12. Linda Ivette Colon Reyes, *Los orígenes de la burguesía y el banco de avío* (Mexico, 1982), 120; and Esteban de Antuñano, *La Industria del Algodón en México* (Puebla, 1833), 43.

13. Mariano Torres Bautista, *Testamento del Administrador* (Puebla, 1989), 7.

14. Rivero Quijano, 1990, 103.

15. Rivero Quijano, 1990, 99.

16. Torres Bautista, 1989, 8.

17. Rivero Quijano, 1990, 111.

18. Miguel Quintana and Fernando Pruneda, "Estudio sobre la modernizacion de la industria nacional textil del algodón . . .", 1943, 24, in AGN, Fondo Gonzalo N. Robles, Caja 76, exp.9.

19. Juan Chavez Orozco, *Monografía Economico-Industrial de la Fabricación de Hilados y Tejidos de Algodón* (Mexico, 1933) 36.

20. Pruneda and Quintana, 1943, p. 25.

21. de Antuñano, 1833, 74.

22. "European and American businesses also hired British workers." Peter N. Stearns, *The Industrial Revolution in World History* (Boulder, 1993), 43.

23. Rivero Quijano, 1990, 122.

24. Pruneda and Quintana, 1943, 24–25.

25. Pruneda and Quintana, 1943, 26.

26. Armando Razo and Stephen Haber, "The Rate of Growth of Productivity in Mexico, 1850–1933," in *The Journal of Latin American Studies 30*, 502.

27. R. A. Buchanan, *The Power of the Machine* (London, 1992), chapter 5.

28. Miguel Casas, Informe, February 10, 1913, AGN, DT, Caja 6, Exp. 27.

29. To Yllmo Señor, November 26, 1877, archivo del Catedral San Miguel Arcangel, Orizaba.

30. Pruneda and Quintana, 1943, 34.

31. The best description of this process is in Stephen Haber, Armando Razo, and Noel Maurer, *The Politics of Property Rights, Political Instability, Credible Commitments, and Economic Growth in Mexico, 1876–1929* (Cambridge, 2004), 382.

32. John H. Coatsworth, *Growth Against Development, The Economic Impact of Railroads in Porfirian Mexico* (DeKalb, 1981), 4.

33. Carmen Ramos Escandon, "Working Class Formation and the Mexican Textile Industry: 1880–1912." Ph.D. diss., State University of New York at Stony Brook, 1981, 45.

34. INEGI, *Estadísticas Históricas de México, 1994*, 616.

35. Stephen H. Haber, *Industry and Underdevelopment, the Industrialization of Mexico, 1890–1940* (Stanford, 1989), 190–93.

36. Razo and Haber, in *The Journal of Latin American Studies 30*, 512.

37. Rodney Anderson, *Outcasts in Their Own Land* (De Kalb, 1976), 100.

38. Juan Chávez Orozco, *Monografía Económico-Industrial de la Fabricación de Hilados y Tejidos de Algodón* (Mexico, 1933), 7 and 8.

39. C. Reginald Enock, *Mexico* (London, 1909), 337.

40. Quijano, 1990, T. I, 105.

41. Sanford Mosk, *Industrial Revolution in Mexico* (Berkeley, 1954), 123.

42. Raquel Beato King, "La Industria Textil Fábril en México, 1830–1910" (Doctorado en Historia Economica, 1977), 104.

43. Leticia Gamboa Ojeda, *Los Empresarios de Ayer, El grupo dominante en la industria textil de Puebla 1906–1929*, (Puebla, 1985), 162.

44. Eugenio y Julio Juan Lions, Luis Moutte, Adrian Reynaud, and Casimiro Philip owned Lions Hermanos. Adrian Reynaud and Leon Signoret owned Signoret y Reynaud.

45. Gamboa, 1985, 166.

46. Quijano, 1990, T. I, 169–82.

47. Stephen Haber and Armando Razo, "Political Instability and Economic Performance: Evidence from Revolutionary Mexico, *World Politics*, Vol. 51, No. 1, October 1998, 132.

48. Haber, 1989, 55.

49. Jeffrey Bortz, "The Cotton Textile Industry," *The Encyclopedia of Mexico* (Chicago, 1997).

50. Chávez Orozco, 1933, 68.

51. Dawn Keremitsis, *La Industria Textil Mexicana en el Siglo XIX* (Mexico, 1973), 113–17.

52. SICT, "Fábricas de Hilados y Tejidos en la República en el año de 1921, April 18, 1922, AGN, DT, Caja 299, Exp. 1.

53. Mario Ramírez Rancaño, *burguesía textil y política en la revolución mexicana* (Mexico, 1987), 20. For an ample discussion of the 1906 textile strikes, see Anderson, 1976, chapter 3, "The Year of the Strikes."

54. Rancaño, 1987, 63–64.

55. Enock, 1909, 132.

56. Haber, Razo, and Maurer, 2004, 345–47.

57. Marcos López Jiménez (El Subdirector, Encargado del Departamento), Exposición de motivos . . . , December 19, 1914, AGN, DT, Caja 89, Exp. 3.

58. Angel Lerdo de Tejada to Director del DT, September 4, 1913, AGN, DT, Caja 47, Exp. 26.

59. Miguel Casas to Adalberto Esteva, November 27, 1913, AGN, DT, Caja 51, Exp. 10.

60. Manuel Díaz to Román Rodríguez Peña, March 21, 1913, AGN, DT, Caja 51, Exp. 1.

61. Compañía Industrial de Atlixco, S.A., Memorandum, October 18, 1928, AGN, DT, Caja 1410, Exp. 5.

62. *El Democrata*, Mexico City, January 4, 1916, 1.

63. *El Dictamen*, January 5, 1916, 1.

64. Galindo and Heredia to Director, November 13, 1914 , AGN, DT, Caja 91, Exp. 8.

65. INEGI, *Estadísticas Históricas de México*, 1999, 303, 507, 517, 519.

66. Gruening, 1928, 349.

67. SICT, *Monografía sobre el Estado Actual de la Industria en México* (Mexico, 1929), 32.

68. SICT, "Fábricas de Hilados y Tejidos en la República en el año de 1921, April 18, 1922, AGN, DT, Caja 299, Exp. 1.

69. SICT, *Monografía sobre el Estado Actual de la Industria en México* (Mexico, 1929), 34–35.

70. Enock, 1909, 185.

71. María del Socorro Quesada Salcedo, *La Evolución de la División Municipal según los Censos de Población* (Mexico, 1960), 51, 52.

72. Nevin O. Winter, *Mexico and Her People of To-Day* (Boston, 1907), 46.

73. Enock, 1909, 189.

74. INEGI, *Estadísticas Históricas de Mexico* (Mexico, 1994), 17.

75. Soñia Pérez Toledo, *Los Hijos del Trabajo* (Mexico, 1996), 259.

76. Carlos Illades, *Hacia la República del Trabajo* (Mexico, 1996), 30.

77. Mario Trujillo Bolio, Operarios Fabriles en el Valle de México (1864–1884) (Mexico, 1997), 30.

78. John Lear, "Workers, Vecinos and Citizens: The Revolution in Mexico City, 1909–1917" (Ph.D. dissertation, UC Berkeley, 1993), 40.

79. Cuestionario para el censo industrial y obrero, La Carolina, March 21, 1921, AGN, DT, Caja 288, Exp. 6.

80. SICT, Cuestionario sobre trabajo, San Antonio Abad, August 23, 1920, AGN, DT, Caja 299, Exp. 1.

81. Cuestionario para el censo industrial y obrero, San Antonio Abad, February 12, 1921, AGN, DT, Caja 288, Exp. 6.

82. Cuestionario para el censo industrial y obrero, La Azteca, April 21, 1921, AGN, DT, Caja 288, Exp. 6.

83. The area no longer pertains to Tlalnepantla but Naucalpan.

84. Visita de inspección, September 23, 1913, AGN, DT, Caja 51, Exp. 14.

85. Secretaría de la Economía Nacional, Departamento de Estudios Económicos, *La Industria Textil en México, El Problema Obrero y Los Problemas Económicos* (Mexico, 1934), 256.

86. Enock, 1909, 320.

87. Bernardo García Díaz, *Textiles del Valle de Orizaba 1880–1925* (Xalapa, 1990), 22.

88. Felipe Franco, *Geografía de Puebla* (Mexico, 1941), 31.

89. INEGI, 1994, 37.

90. INEGI, 1994, 24.

91. Manuel Ortega Elorza, June 14, 1913, AGN, DT, Caja 51, Exp. 15.

92. Franco, 1941, 70–72.

93. Franco, 1941, 34–36.

94. Samuel Malpica, *Atlixco: Historia de la Clase Obrera* (Puebla, 1989), p. 49.

95. A history of the Maurer family can be found in Mariano Torres Bautista, *La familia Maurer de Atlixco, Puebla* (Mexico, 1994)

96. Malpica, 1989, p. 50.

97. Malpica, 1989, p. 51.

98. Malpica, 1989, pp. 52–53.

99. Malpica, 1989, pp. 55–56.

100. Benito Flores, April 8, 1924, Presidencia, 1924, num. H3.

101. Leonardo Altamirano to Presidente de la Junta Central de Conciliación y Arbitraje, September 30, 1932, AMO, Caja 913, exp. 41.

102. Ibid.

103. García Díaz, 1990, 17.

104. Winter, 1907, 93.

105. Leonardo Altamirano to Presidente de la Junta Central de Conciliación y Arbitraje, September 30, 1932, AMO, Caja 913, exp. 41, 11.

106. Roberto Saviñon, Inspector, Principio del informe . . . , February 23, 1922, AGN, DT, Caja 453, Exp. 9.

107. Peake, 34.

108. Peake, 37.

109. Peake, 47.

110. Peake, 64.

111. Peake, 82.

112. Peake, 86.

113. Miguel Quintana, "Estudio sobre la modernación . . .", in AGN, Gonzalo N. Robles, Caja 76, Exp. 9.

114. Rivero Quijano, 81.
115. Antonio de Zamacona to Señor Director, May 27, 1912, AGN, DT, Caja 7, Exp. 13.

CHAPTER THREE

1. Marcelino Duran, Eusebio García, and Juan Villegas to Presidente de la Junta de Administración Civil, April 16, 1928, AMO, Caja 201, Exp. 1.
2. Copia Integra de las Declaraciones Rendidas en la Oficina de la Policia Judicial . . .", April 12, 1928, AMO, Caja 201, Exp. 1.
3. El Oficial Judicial to Presidente de la Junta de Administración Civil, May 2, 1928, AMO, Caja 201, Exp. 1.
4. Fernando Noriega et al. to Comité de Obreros, January 3, 1914, AGN, DT, Caja 52, Exp. 3.
5. Rodolfo Flores Gama and Carlos Maldonado to Marcos López Jiménez, June 1915, AGN, DT, Caja 103, Exp. 9.
6. Florencio Pastrana to Marcos López Jiménez, May 26, 1915, AGN, DT, Caja 99, Exp. 10.
7. Enrique Hinojosa, et al., January 13, 1915, AGN, DT, Caja 104, Exp. 6.
8. Srio Gral et al. to Presidente Municipal, February 28, 1919, AMA, Presidencia, 1919. Author's underline.
9. Nevin O. Winter, *Mexico and Her People of To-Day* (Boston, 1910), 53.
10. C. Reginold Enock, *Mexico: Its Ancient and Modern Civilisation* (London, 1909), 167.
11. Leandro Aguilar y Sánchez, et al, "A los Obreros . . . ," November 29, 1907, AMA, Presidencia, Gobernación, 1907, Caja 164.
12. "Los humildes Servidores de la Fábrica la Guia" to Ramos Pedrueza, January 13, 1913, AGN, DT, Caja 52, Exp. 1.
13. "un obrero del estampado" to Director del Departamento del Trabajo, May 18, 1913, AGN, DT, Caja 52, Exp. 1.
14. "Los obreros de San Antonio Abad", June 2, 1913, AGN, DT, Caja 52, Exp. 1.
15. "Barios Obreros" to "Director del Departamento de Trabajo, May 27, 1913, AGN, DT, Caja 52, Exp. 1.
16. Fábrica de Metepec, July 26, 1913, AGN, DT, Caja 52, Exp. 1.
17. Entrevista, March 21, 1915, AGN, DT, Caja 106, Exp. 20.
18. Srio Gral et al. to Srio Gral de la Fed. Puebla, November 14, 1917, ASSR 6, 1917, Exp. 611.09 (8).
19. Antonio Valero to Oficial Mayor de la Secretaría de Fomento, September 15, 1914, AGN, DT, Caja 87, Exp. 13.
20. Gabriel Domingues, Tehuacan, Círcular Num. 12, April 21, 1926, Archivo Crom-Atlixco, SG, 1926
21. Jose Otáñez to Antonio Ramos Pedrueza, February 14, 1912, AGN, DT, Caja 7, Exp. 18.
22. Adalberto Esteva, "Labor del Departamento del Trabajo." *Boletín del Departamento del Trabajo,* July 1913, Año 1, Num. 1, 2.

23. DT to Julio Vivar, May 1, 1913, AGN, DT, Caja 37, Exp. 40.

24. "Reglamento," in Jésus Rivero Quijano, *La Revolución Industrial y La Industria Textil en México* (Mexico, 1990) , 137.

25. *Código del Trabajo Expedido por el XXVI Congreso Constitucional del Estado de Campeche* (Campeche, 1918), 4.

26. Constitución Politica, Artículo 123, 39.

27. Constitución Politica, Artículo 123, Section XX, 41.

28. Constitución Politica, Artículo 123, Section XXII, 41.

29. *Ley del Trabajo del Estado Libre y Soberano de Veracruz-Llave* (Orizaba, 1918), 59.

30. Vicente Yslas González et al. to Cámara de Trabajo de Atlixco, July 3, 1926, Archivo Crom-Atlixco, SG, 1926.

31. Epitacio López to Cámara de Trabajo de Atlixco, June 22, 1926, Archivo Crom-Atlixco, SG, 1926.

32. Guillermo Palacios to Enrique César, June 31, 1933, AMO, Caja 917, Exp. 64.

33. Luis Osorio, n.d. (probably 1916), ASSR, 1916–1918.

34. SICT, Cuestionario sobre trabajo, February 3, 1920, AGN, DT, Caja 207, Exp. 20.

35. SICT, Cuestionario sobre trabajo, March 6, 1920, AGN, DT, Caja 207, Exp. 26.

36. SICT, Cuestionario sobre trabajo, March 6, 1920, AGN, DT, Caja 207, Exp. 26.

37. Informe de visita de inspección practicada a la Pasamaneria Francesa S.A., AGN, DT, Caja 223, Exp. 6.

38. SICT, Cuestionario, December 1922, AGN, DT, Caja 405, Exp. 1.

39. *El Imparcial*, 1/27/12, p. 1

40. The data in La Carolina is skewed by the mill's second largest job category, "diversos." The 347 workers in this category earned an average wage of 2.00 pesos and perhaps included both skilled and unskilled workers.

41. "En la Villa de Río Blanco . . . ," March 10, 1916, AGN, DT, ASRB, 1916–18.

42. Director to Señores Representantes, April 9, 1920, ASRB, Caja 1912–1920.

43. *Boletín del Departamento del Trabajo,* January 1914, Año I, Núm. 7, 620–623.

44. Antonio Ramos Pedrueza, sesión del día July 8, 1912, AGN, DT, Caja 15, Exp. 18.

45. DT, Cuestionario sobre Trabajo, February 3, 1920, Caja 207, Exp. 20; SICT, Lista de Industrias, January 25, 1924, AGN, DT, Caja 279, Exp. 5.

46. DT, Cuestionario de huelgas, August 1923, Caja 558, Exp. 9.

47. El Gobernador del Estado, February 10, 1923, AGN, DT, Caja 558, Exp. 4; Caja 1171, Exp. 1; Dept. de Trabajo, Cuestionario de huelgas, January, 1923, Caja 558, Exp. 9.

48. Huelgas, May 1923, AMA, Presidencia, 1923, Caja H-19, num. 9; Huelgas, n.d., AMA, Presidencia, 1922, Caja B-11, Exp. Huelgas; Huelgas, January 1923, AMA, Presidencia, 1923, Caja H-19, Núm. 9; Cuestionario de Accidentes, 1923, AMA, Presidencia, 1923, Caja H-9, Núm. 9.

49. *Boletín de Industria, Comercio y Trabajo,* July, August, September 1920, Sección de Estadística, "Huelgas en la República Abril 1920."

50. Janet Hunter and Helen Macnaughtan, "Gender and the Global Textile Industry, 1650–2000," paper presented to the IISH textile conference, Amsterdam, Nov. 11–13, 2004.

51. Peake, n.d., 72,75, 79, 90.

52. Peter N. Stearns, *The Industrial Revolution in World History* (Boulder, 1993), 61.

53. L. Folsa Vda de Menual, July 17, 1914, AGN, DT, Caja 91, Exp. 3.

54. Margarita Flores to Martín Torres, August 11, 1932, AMO, Caja 917.

55. She wrote the letter because a policeman arrested her when she had rejected his *"torpes pretensiones, requiriendome de amores."*

56. Rosario Molina Parra, "A Orgullo Tengo 87 años," *Los Días Eran Nuestros,* 71.

57. Otilia Zumaya, July 16, 1914, AGN, DT, Caja 91, Exp. 3.

58. Adela G. Viuda de Ysassi, July 17, 1914, AGN, DT, Caja 91, Exp. 3.

59. Ma. Ana E. Viuda de Morelos, July 17, 1914, AGN, DT, Caja 91, Exp. 3.

60. Otilia Zumaya, July 16, 1914, AGN, DT, Caja 91, Exp. 3.

61. Rosario Rodríguez, July 17, 1914, AGN, DT, Caja 91, Exp. 3.

62. Lear, 1993, 32.

63. Manuel R. Díaz to Señor Director del Departamento del Trabajo, February 8, 1913, AGN, DT, Caja 51, Exp. 1.

64. W. O. Jenkins to Antonio Ramos Pedrueza, March 2, 1912, AGN, DT, Caja 7, Exp. 20.

65. "Resumen del personal . . . ," April 21, 1926, AGN, DT, Caja 976, Exp. 13.

66. Ibid.

67. Manuel Díaz, informe, February 4, 1913, AGN, DT, Caja 51, Exp. 1.

68. W. O. Jenkins to Antonio Ramos Pedrueza, March 26, 1912, AGN, DT, Caja 7, Exp. 20.

69. Alcalde Municipal, January 9, 1904, AMCM, 1903–1909, Exp. 4.

70. Francisco Reyes López to Martín Torres, October 14, 1932, AMO, Caja 917, Exp. 62.

71. *Boletín de Industria, Comercio y Trabajo,* July, August, September 1920, Sección de Estadistica, "Huelgas en la República Abril 1920."

72. Cuestionario, September 1922, AGN, DT, Caja 405, Exp. 1.

73. Francisco Reyes López to Martín Torres, October 14, 1932, AMO, Caja 917, Exp. 62.

74. Fábrica La Esperanza, Cuestionario de Salarios, June 1921, AGN, DT, Caja 299, Exp. 1.

75. Maurilio Luna et al., October 4, 1917, ASSR, 7, 1917, 714.1.

76. INEGI, *Estadísticas Históricas de México,* 1994, 348.

77. Leticia Gamboa, "Los Obreros Textiles de Puebla. El Caso de Atlixco, 1899–1924. Ph.D. thesis, Universidad de Paris VIII, 1993, 111.

78. Gamboa, 1993, 117.

79. John Lear, "Workers, Vecinos and Citizens: The Revolution in Mexico City, 1909–1917" (Ph.D. dissertation, UC Berkeley, 1993), 18.

80. John Lear, 1993, 19–20.

81. Bernardo García Díaz, *Textiles del Valle de Orizaba 1880–1925* (Xalapa, 1990), 61–62.

82. Miguel Casas to Adalberto Esteva, November 28, 1913, AGN, DT, Caja 51, Exp. 10.

83. Alan Knight, *The Mexican Revolution* (Lincoln, 1990), vol. 1, 134.

84. Roberto Saviñón to Jefe del Departamento del Trabajo, December 30, 1921, AGN, DT, Caja 325, Exp. 12.

85. la empresa construyó para los obreros una serie uniforme y en fila de cuartos de adobe, con techos de dos aguas y dos puertas, alrededor de las galeras rectangulares colocadas sobre dos avenidas, dejando en el centro un patio con un tanque para recolectar el agua de los lavaderos y las letrinas colectivas. Estas manzanas parecen copiadas de los cottages ingleses descritos por Engels; este caserío formaba una verdadera Company Town. Para transitar por el se les exigía a los obreros una tarjeta de identificación, so pena de ser encarcelados en el cuartel de los rurales federales que se encontraba en el interior suroccidental de la fábrica y multados por el juez auxiliar. Malpica, 1989, 59.

86. SICT, Cuestionario sobre trabajo, June 6, 1921, AGN, DT, Caja 299, Exp. 1.

87. Manuel Ortega Elorza to Director, September 23, 1913, AGN, DT, Caja 51, Exp. 14.

88. Manuel Ortega Elorza, visita de inspección, AGN, DT, Caja 51, Exp. 15.

89. Martínez Carrillo, "Sesión del Dia 8 de Julio de 1912, Convención de Industriales," AGN, DT, Caja 15, Exp. 18.

90. M. R. Díaz, March 22, 1915, AGN, DT, Caja 97, Exp. 30.

91. J. de Beraza to Julio Poulat, June 1, 1920, AGN, DT, Caja 223, Exp. 10.

92. Manuel Aparicio Guido, *Resumenes de Sociología y de Economía Social* (Jalapa, 1923), 144.

93. Leonardo Altamirano to Presidente de la Junta Central de Conciliación y Arbitraje, September 30, 1932, AMO, Caja 913, Exp. 41.

94. Adela Vda de Ysassi to Rafael Salas, July 17, 1914, AGN, DT, Caja 91, Exp. 3.

95. Pánfilo Méndez et al. to Alcalde Municipal, December 26, 1908, ASRB, 1912–1920.

96. E. J. Hernández to Honorable Consejo Federal, February 20, 1927, Archivo CROM Atlixco, SG, 1927.

97. Brigido Flórez to Ricardo Carreto, Archivo CROM Atlixco, SG, 1926.

98. Leopoldo Lezama to Juan Pérez y Pérez, Srio Gral de la Cámara del Trabajo, Atlixco, February 21, 1927, Archivo CROM Atlixco, SG, 1927.

99. Entrevista, March 21, 1915, AGN, DT, Caja 106, Exp. 20.

100. Acta del Sindicato, 6/14/16, AGEV, DEPS, 1916, Exp. 703 (21-c).

101. Cuestionario . . . July 21, 1921, AGN, DT, Caja 299, Exp. 1.

102. J. de Beraza to Julio Poulat, June 1, 1920, AGN, DT, Caja 223, Exp. 10.

103. Facundo Pérez Linares, "Los años que fueron nuestros," *Los Días Eran Nuestros,* Puebla, 1988, 164.

104. AGN, DT, Caja 106, Exp. 20, entrevista, March 21, 1915.

105. Entrevista celebrada entre la Comisión del Centro Fabril de Orizaba y el C. Ministro de Gobernación, Lic. Rafael Zubaran Capmany, March 21, 1915, AGN, DT, Caja 106, Exp. 20.

106. "Horas de entrada y Salida al Trabajo," AGN, DT, Caja 979, Exp. 3.

107. *El Imparcial,* January 14, 1912, 10.

108. Carlota Vda. de Ortega, July 17, 1914, AGN, DT, Caja 91, Exp. 3.

109. Adela G. Vda. de Ysassi, July 23, 1914, AGN, DT, Caja 91, Exp. 4.

110. Elena de la Vega, July 20, 1914, AGN, DT, Caja 91, Exp. 4.

111. Otilia Zumaya, July 19, 1914, AGN, DT, Caja 91, Exp. 4.

112. Antonio de Zamacona to Rafael L. Hernández, November 14, 1912, AGN, DT, Caja 3, Exp. 12, 1912. Antonio Ramos Pedrueza to Manuel Rivero Collado, March 20, 1912, AGN, DT, Caja 7, Exp. 19.

113. Memorandum Num. 1., n.d., AGN, DT, Caja 1159, Exp. 2.

114. Islas, 1956, 130, 131.

115. Asunto Toquero, July 11, 1916, ASRB, folder 1916–1918.

116. Chester Jones, *Mexico and its Reconstruction* (New York, 1921), 108.

117. Esteban de Antuñano, *La Industria del Algodón en México* (Puebla, 1833), 66.

118. José Fernandez Rojas, *La Revolución Mexicana de Porfirio Díaz a Victoriano Huerta 1910–1913* (Mexico, 1913), 96.

119. Francisco Escudero, "El regimen parlamentario convendría a México dada su composición social y su actual desarrollo político?" in *Boletín de la Secretaría de Gobernación,* Vol. I, Num. 2, Mexico D.F. July 1922, 519.

120. Winter, 1910, 184–185.

121. Jones, 1921, 109.

122. Jones, 1921, 112.

123. Gamboa Ojeda, Leticia, *Los Empresarios de Ayer, El grupo dominante en la industria textil de Puebla 1906–1929,* (Puebla, 1985), 164.

124. *El Imparcial,* January 6, 1912, p. 1.

125. Bernardo Barrolo, administrador, El León, to Juez de Paz de esta fábrica, August 14, 1920, AMA, Presidencia, 1920, Caja 413.

126. Vicente Lombardo Toledano, *La Libertad Sindical en México* (Mexico, 1926), 195.

127. El Comité Ejecutivo to Jefe del Departamento de Fomento y Agricultura, June 17, 1917, ASSR, 7, 1916–1918.

128. Tomas Reyes Retana, "Sesión del Día 8 de Julio de 1912," Convención de Industriales, AGN, DT, Caja 15, Exp. 18.

129. November 26, 1877, Archivo Catedral San Miguel Arcangel, Orizaba, Veracruz.

130. Srio de Gobierno to Pres. Municipal de Orizaba, January 28, 1920, AMO, Caja No. 634, Exp. 133.

131. Juan Olivares, Luis Lopez, Gilberto Ruiz to A. Ramos Pedrueza, April 18, 1912 AGN, DT, Caja 7, Exp. 12.

132. Miguel Casas to Adalberto Esteva, March 1914. AGN, DT, Caja 51, Exp. 14.

133. José Natividad Díaz, Presidente, and Francisco de B. Salazar, Secretario, to Director, Depto. de Trabajo, January 8, 1914, AGN, DT, Caja 91, Exp. 19.

134. Pablo Méndez, March 21, 1923, AGEV, TPS, 1923 (3854), num. 183.

135. Vázquez to las Agrupaciones Hermanas, May 14, 1927, Archivo CROM Atlixco, Sría Gral, 1927.

136. Srio Gral to Miguel Hidalgo, February 17, 1927, AMA, Presidencia, 1927, num. 7.

137. Srio Gral to Miguel Hidalgo, March 1, 1927, AMA, Presidencia, 1927, num. 7.

138. Obreros to Presidente Municipal, January 4, 1927, AMA, Presidencia, 1927, num. 7.

139. J. C. Valadez to D. Abad de Santillan, March 16, 1924, Abad de Santillan, International Institute of Social History, Amsterdam.

140. Enock, 1909, 238.

141. Carlos Aldeco, a los habitantes de esta ciudad . . . , AGMP, Series de Expedientes Caja 4, Exp. 49.

142. Pulque is a fermented juice of the maguey and a traditional drink in Mesoamerica. It is an acquired taste.

143. Family Spending Surveys 1914, AGN, DT, Caja 68, Exp. 1.

144. Facundo Pérez Linares, "Los años que fueron nuestros," *Los Días Eran Nuestros*, Puebla, 1988, 173.

145. *"La cucaracha, la cucaracha, ya no puede caminar, porque no tiene, porque le falta, marijuana que fumar."*

146. Rivero Quijano, 1990, 399.

147. Francisco Reyes López to Martín Torres, October 14, 1932, AMO, Caja 917, Exp. 62.

148. "Combenio celebrado," January 1920, ASRB, folder 1908–1920.

149. Jesús Loaiza to Srio Gral de la Cámara del Trabajo, October 6, 1925 Archivo Crom-Atlixco, SG, 1925.

150. Ernest Gruening, *Mexico and Its Heritage* (New York, 1928), 250.

151. Gruening, 1928, 250.

152. Gruening, 1928, 250.

153. Municipio de Puebla, Reglamento Para la Prostitución en Puebla de Zaragoza, AGMP, Series de Expedientes, Caja 1, Exp. 14, 1902.

154. Municipio de Puebla, Reglamento Interior de la Sección de Sanidad, AGMP, Series de Expedientes 1915, Caja 6, Exp. 82.

155. Juan Pérez et al. to H. Ayuntamiento, March 3, 1920, AMO, Caja No. 636, Exp. 229.

156. Luis Lara y Pardo, *La Prostitución en México* (Mexico City, 1908), 19.

157. Jose Lupercio, July 19, 1920, AMO, Caja No. 640, Exp. 436.

158. Memorandum sobre Reglamentación de la Prostitución, June 12, 1947, AGN, DGG, 2.331, Caja 79A, Exp. 70.

159. Roberto López, Inspector de la Ley del Trabajo, 1916, AGEV, DEPS, 1916, Exp. 756 (74-s).

160. *El Universal*, May 4, 1917, 8.

161. Pro-Paría, March 14, 1920, Año III, Tomo III, Num. 80.

162. Francisco Reyes López to Martín Torres, October 14, 1932, AMO, Caja 917, Exp. 62.

163. Francisco Reyes López to Martín Torres, October 14, 1932, AMO, Caja 917, Exp. 62.

164. Raymond Williams, *The Long Revolution* (London, 1961), 72.

165. John M. Hart, *Anarchism and the Mexican Working Class, 1860–1931* (Austin, 1987).

CHAPTER FOUR

1. While some historians argue that Carranza's defeat of the 1916 Casa del Obrero Mundial (COM) general strike suggest a state strong enough to defeat labor, the data suggest it was not a defeat for textile workers and certainly not for the growing workers and union movement in Mexico.

2. The Mexican state came to control labor in the 1920s, tamed it in the 1930s and 1940s, and began to crush it with the railroad strike of 1958. These three stories are best left for another volume.

3. Mariano Torres Bautista, *Testamento del Administrador* (Puebla, 1989), 9–17.

4. Carlos Illades, *Hacia la República del Trabajo* (Mexico, 1996), 154.

5. Sonia Perez Toledo, *Los Hijos del Trabajo* (Mexico,1996), 244.

6. Mario Trujillo Bolillo, *Operarios Fabriles en el Valle de México* (Mexico, 1997), 266.

7. Unión de veteranos del trabajo, *Sucesos del Trabajo y sus Luchas de Antaño. Obra que narra los sufrimientos del trabajador desde 1800 a 1907* (Puebla, 1938), 111.

8. Moises Gonzalez Navarro, *Las Huelgas Textiles en el Porfiriato* (Puebla, 1970), 13.

9. González Navarro, 1970, 15–42.

10. David Walker, "Porfirian Labor Politics: Working Class Organization in Mexico City and Porfirio Díaz, 1876–1902," *The Americas,* vol. XXXVII, no. 3, January 1981, 282.

11. Blanca Estela Santibañez Tijerina, "Las Huelgas Textiles Tlaxcaltecas Durante el Porfiriato" (Puebla, 1996), 22.

12. A copy of the laudo is in AGN, Gobernación-legagos, Caja 817, Exp. 1, Correspondencia Particular del Ministro de Gobernación, n.d.; and an English translation in Karl Koth, "Not a Mutiny but a Revolution: The Río Blanco Labour Dispute, 1906–1907," *Canadian Journal of Latin American and Caribbean Studies,* vol. 18, no. 35 (1993), 57–58.

13. Hart, 1987, 97.

14. Koth, 1993, 49.

15. Pascual Mendoza, Adolfo Ramirez, Antonio Espinosa, Reglamento, December 9, 1906, AGN, Gobernación-Legajos, Caja 817, Exp. 1.

16. Author's underline. Pascual Mendoza, Adolfo Ramirez, Antonio Espinosa, Reglamento, December 9, 1906, AGN, Gobernación-Legajos, Caja 817, Exp. 1.

17. Leticia Gamboa Ojeda, "La Huelga Textil de 1906–1907 en Atlixco," *Historia Mexicana,* Vol XLI, Num. 1, July–September 1991, 153.

18. John Mason Hart, *Revolutionary Mexico* (Berkeley, 1987), 98.

19. Author's underline. Manuel Rivero Collada to Ramón Corral, January 6, 1907, AGN, Gobernación-legagos, Caja 817, Exp. 1.

20. Informes de Orizaba, November 12, 1906, AGN, Gobernación-legagos, Caja 817, Exp. 1.

21. Samuel Malpica, "Anarcosindicalismo o sindicalismo revolucionario en México" in Jaime Tamayo and Patricia Valles, *Anarquismo, socialismo y sindicalismo en las regiones* (Guadalajara, 1993), 93.

22. AMA, Presidencia, 1910, Gobernación, Sindicatos, Caja 246.

23. Hart, 1987, 99–100.

24. Hart, 1997, 72.

25. Modesto Tinoco, et al. to Señor Administrador, May 10, 1909, AMA, Presidencia, Gobernación, Caja 211, 1909.

26. To Sría. Gral. del Departamento Ejecutivo del Estado, May 13, 1909, AMA, Presidencia, Gobernación, Caja 211, 1909.

27. Cía Ind de Atlixco to Ignacio Machorro, May 17, 1909, AMA, Presidencia, Gobernación Caja 211, 1909.

28. La Comición to Mateo Arcos, January 22, 1910. AMA, Presidencia, Gobernación, 1910, Caja 246.

29. To Jefe Político del Distrito, February 3, 1910; José Nieto, Braulio Cortes, Toribio Cortes y Cabrera, Manuel García, Bernardino Jacome, Genaro Gutiérrez, Justino Mosqueira, Joaquin Morales to Ignacio Machorro, February 8, 1910; Ignacio Machorro, April 4, 1910, AMA, Presidencia 1910, Caja 246.

30. Ignacio Machorro, April 4, 1910, AMA, Presidencia 1910, Caja 246.

31. Pedro Luna et al. to Gerente y Administrador de esta Fábrica, October 3, 1910, AMA, Presidencia 1910, Caja 246.

32. Hart, 1987, 92.

33. Hart, 1987, 91.

34. Alan Knight, *The Mexican Revolution* (Lincoln, 1990), vol. 1, 127–140.

35. David LaFrance, *The Mexican Revolution in Puebla, 1908–1913* (Wilmington, 1989), 4.

36. LaFrance, 1989, 65.

37. LaFrance, 1989, 80.

38. Knight, 1990, vol. 1, 238.

39. Juan Felipe Leal, "Las Agrupaciones Obreras (1911–1913)" in Juan Felipe Leal and José Villasenor, *en la revolución 1910–1917* (Mexico, 1988), 167–68.

40. Gutierrez, 2000, 253.

41. Díaz Caneja to Ygnacio Machorro, Metepec, January 19, 1911; to Jefe Político del Distrito, Atlixco, January 11, 1911; Vicente Velázquez to Jefe Político de Atlixco, Metepec, February 20, 1911; Vicente Velázquez to Jefe Político de Atlixco, Metepec, March 13, 1911; AMA, Presidencia, 1911, Caja H-2.

42. E.A.M.P.M.D.L.L., September 6, 1911, AMO, 1911.

43. E.A.M.P.M.D.L.L. to Juez 1 de Paz, September 6 and 7, 1911; E.A.M.P.M.D.L.L. to Teniente Coronel del 15 Batallon, September 6, 1911, AMO, 1911.

44. E.A.M.P.M.D.L.L. to Jefe Politico del Canton, September 11, 1911, AMO, 1911.

45. Subdirector to Juan Camacho, September 12, 1911, AMO, 1911.

46. *El Imparcial*, January 21, 1912.

47. *El Imparcial*, December 22, 1911.

48. *El Imparcial*, December 11, 1911, 1.

49. *El Imparcial*, December 28, 1911, 1.

50. *El Imparcial*, January 14, 1912, 10.

51. *El Imparcial*, December 28, 1911, 10, 11; December 29, 1911, 1.

52. Nicolas Meléndez, "Contestación del C. Gobernador al Presidente del Congreso durante la protesta otorgada el dia 25 de Diciembre de 1911," AGMP, Series de Expedientes, Caja 3, Exp. 38.

53. *El Imparcial*, December 29, 1911, 8.

54. *El Imparcial*, December 31, 1911, 1.

55. *"los 'leaders' del movimiento se inclinan ya hacia una actitud conciliadora, el resto de huelguistas dice que no se conformara con que solo se le concedan menos horas de labor y nada de aumento a los salarios, manifestando que prefieren ganar mas, aunque el trabajo sea igual."* *El Imparcial*, December 29, 1911, 8.

56. *El Imparcial*, December 29, 1911, 8.

57. *El Imparcial*, January 5, 1912, 1.

58. Jorge Durand, *Los obreros de Río Grande* (Michoacan, 1986), 77.

59. *El Imparcial*, January 6, 1912, 1.

60. *El Imparcial*, January 6, 1912, 1.

61. *El Imparcial*, January 6/1912, 1.

62. *El Imparcial*, January 6/1912, 1.

63. Francisco Romero Sánchez, Pánfilo Méndez, Ramón G. Camarillo, Daniel Tejada, Manuel Velasco, José Mendoza, Lucio Alvarado, Lauro Luna, Leobardo F. Huerta. *El Imparcial*, January 16, 1912, 1, 6.

64. To Secretario del Gobierno del Estado, January 19, 1912, AGEV, Sección de Fomento, Caja 27, Exp. 4s, 1912; *El Imparcial*, January 9, 1912, 1.

65. "Como se estableció la tarifa mínima de salarios para los obreros de hilados y tejidos de algodón, *Boletín del Departamento del Trabajo*, Año 1, no. 1, July 1913, 21.

66. "El Primer Paso," *Boletín del Departamento del Trabajo*, Año I, no. 2, August 1913, 108.

67. Rafael Hernández, Srio de Fomento, to Gobernador del Estado de Veracruz, January 29, 1912, Archivo General del Estado de Veracruz (AGEV), Sección de Fomento, Caja 27, Exp. 4s, 1912.

68. *El Imparcial*, January 21, 1912, 1, 10. "El Primer Paso," *Boletín del Departamento del Trabajo*, Año I, no. 2, August 1913, 108.

69. *El Imparcial*, January 31, 1912, 1.

70. *El Imparcial*, February 10, 1912, 1.

71. Ponton to Jefe Político, January 23, 1912; Centro Industrial Mexicano to Secretario de Gobernación, February 5, 1912; Díaz Cañeja to Jefe Político, September 19, 1912, AMA, Presidencia, Gobernación, 1912, Caja 301.

72. Jesús Burgos to Jefe Político, July 22, 1912; Arturo Cid to Jefe Político, October 3, 1912, AMA, Presidencia, Gobernación, 1912, Caja 301.

73. Secretaría de Fomento, Colonización e Industria, Departamento del Trabajo, *Iniciativa y Decreto para su Creación* (Mexico, 1912), 3, 15.

74. Secretaría de Fomento, 1912, 18.

75. Mario Ramírez Rancaño, *Burguesía textil y política en la revolución mexicana* (Mexico, 1987), 34–37.

76. Ramos Pedrueza himself linked the establishment of the Labor Department to the general strike. Antonio Ramos Pedrueza, "A los obreros de las Fábricas de Hilados y Tejidos de la República," August 4, 1912, AGN, DT, Caja 50, Exp. 25. On the budget, *El Subsecretario,* September 27, 1912, AGN, DT, Caja 4, Exp. 8.

77. Ramos Pedrueza, August 4, 1912.

78. Junta, May 6, 1912, AGN, DT, Caja 15, Exp. 18.

79. Marcos López Jiménez, Exposición de Motivos . . . , December 19, 1914, AGN, DT, Caja 89, Exp. 3.

80. Ramírez Rancaño, 1987, 60–61.

81. Ramírez Rancaño, 1987, 63.

82. The group included Adolfo Prieto, Félix Martino, Hipolito Gerard, Manuel Rivero Collada, Guillermo Obregon, Antonio Reynaud, Eusebio Gonzáles, Carlos B. Zetina, Daniel Ituarte, Jacobo Grandison.

83. The two competing owners groups continued to fight this issue for the next thirty years. Jeffrey Bortz, "The Genesis of Mexico's Modern Labor Regime: the 1937–1939 Cotton Textile Convention," *The Americas* 52:1 (July 1995), 43–69.

84. Minutes, May 27, 1912, AGN, DT, Caja 15, Exp. 18.

85. This of course is what a new Mexican government would do for many decades following the revolution.

86. Junta, May 13, 1912; "Observaciones y refutación de la proposición hecha por el Sr. Dn. Gustavo Madero . . . ," May 31, 1912, AGN, DT, Caja 15, Exp. 18.

87. Carmen Ramos Escandon, "Working Class Formation and the Mexican Textile Industry: 1880–1912." Ph.D. diss., State University of New York at Stony Brook, 1981, 233.

88. "El Primer Paso," *Boletín del Departamento del Trabajo,* Año I, no. 3, September 1913, 211.

89. Benjamin Meza, Rafael Silva, et al. to Antonio Ramos Pedrueza, June 6, 1912, AGN, DT, Caja 16, Exp. 3; Ramos Escandon, 1981, 233–40.

90. *El Imparcial,* July 2, 1912, 1.

91. Director del Departamento del Trabajo to Nicolás Meléndez, Gobernador del Estado de Puebla, March 15, 1912, AGN, DT, Caja 7, Exp. 19.

92. Ibid.

93. Ibid.

94. Antonio Ramos Pedrueza to Manuel Rivero Collado, March 20, 1912, AGN, DT, Caja 7, Exp. 19.

95. Miguel López Fuente to Director de la Oficina del Trabajo, March 9, 1912, AGN, DT, Caja 7, Exp. 11.

96. *El Imparcial,* June 16, 1912, 1.

97. *El Imparcial,* June 16, 1912, 1; June 21, 1912, 1, 8.

98. *El Imparcial,* July 3, 1912, 1, 5; July 14, 1.

99. *El Imparcial,* July 5, 1912, 1.

100. *El Imparcial,* July 10, 1912, 1.

101. Antonio Ramos Pedrueza, "A los obreros de las Fábricas de Hilados y Tejidos de la Republica," August 4, 1912, AGN, DT, Caja 50, Exp. 25.

102. "Lista de los delegados . . . ," AGN, DT, Caja 15, Exp. 18. Cotton factories dominated the meeting but other factories had representation. Reyes Retana said that there were 133 textile mills in Mexico, 116 of which fabricated cotton. Sesión del July 2, 1912, AGN, DT, Caja 15, Exp. 18.

103. "Proyecto de Reglamento que deberá regir las sesiones . . . ," June 29, 1912, AGN, DT, Caja 15, Exp. 18.

104. Article 17. "Proyecto de Reglamento que deberá regir las sesiones . . . ," June 29, 1912, AGN, DT, Caja 15, Exp. 18.

105. Sesión, July 2, 1912, AGN, DT, Caja 15, Exp. 18

106. Sesión July 3, 1912, AGN, DT, Caja 15, Exp. 18.

107. Miguel López Fuentes (Secretario, Centro Industrial Mexicano) to Director de la Oficina del Trabajo, July 4, 1912, AGN, Caja 7, Exp. 28.

108. José González Soto y Hno. to director del Departamento del Trabajo, August 10, 1912, AGN, Caja 7, Exp. 28.

109. Telegrama No. 21. Hernández to Sr. Don Nicolás Meléndez, AGN, Caja 7, Exp. 28.

110. "El Primer Paso," *Boletín del Departamento del Trabajo,* Año I, Num. 6, December 1913, 523.

111. Sesión July 15, 1912, AGN, DT, Caja 15, Exp. 18. *El Imparcial,* July 18, 1912, 8.

112. "El Primer Paso," December 1913, 523.

113. *El Imparcial,* July 20, 1912, 1.

114. Sesión July 15, 1912, AGN, DT, Caja 15, Exp. 18.

115. "El Primer Paso," *Boletín del Departamento del Trabajo,* Año I, Num. 6, December 1913, 516.

116. Ibid, 514.

117. Ibid, 516.

118. Sesión July 16, 1912, AGN, DT, Caja 15, Exp. 18.

119. Marjorie Ruth Clark, *Organized Labor in Mexico* (Chapel Hill, 1934), 21.

120. "El Primer Paso," *Boletín del Departamento del Trabajo,* Año I, no. 7, January 1914, 624.

121. Sesión July 24, 1912, AGN, DT, Caja 15, Exp. 18; *El Imparcial,* July 18, 1912, 8.

122. J. Rivero Quijano, *La Industria Textil del Algodón y el Maquinismo en México* (Mexico, 1990), 96.

123. Sesión, July 25, 1912, AGN, DT, Caja 15, Exp. 18.

124. Sesión, July 29, AGN, DT, Caja 15, Exp. 18.

125. Ibid.

126. Ibid.

127. Ibid.

128. Secretaría de Fomento, *Tarifa Mínima Uniforme para las Fábricas de Hilados y Tejidos de Algodón de la República, Aprobada por la Convención de Industriales el 31 de julio de 1912* (Mexico, 1912), 4. A copy can be found in AGN, DT, Caja 17, Exp. 6.

129. Ramos Pedrueza to Obreros de la Fábrica, Circular, January 2, 1913, AGN, DT, Caja 47, Exp. 7.

130. Adalbero A. Esteva, Circular a los obreros, n.d., AGN, DT, Caja 47, Exp. 7.

131. Sesión August 1, 1912, AGN, DT, Caja 15, Exp. 18.

132. *El Imparcial,* August 2, 1912, 8.

133. *El Imparcial,* August 6, 1912, 2.

134. *El Imparcial,* August 2, 1912, 8.

135. *El Imparcial,* July 24, 1912, 1 and 8.

136. Acta de la Sesión de clausura, August 1, 1912, AGN, DT, Caja 15, Exp. 18.

137. *El Imparcial,* July 24, 1912, 1 and 8.

138. *El Imparcial,* July 26, 1912, 1.

139. *El Imparcial,* August 1, 1912, 1.

140. Sesión de Clausura, August 1, 1912, AGN, DT, Caja 15, Exp. 18.

141. *El Imparcial,* August 2, 1912, 1; Director del Departamento to Señor Jefe Político del Canton, January 7, 1913, AGN, DT, Caja 54, Exp. 2. The Labor Office also advised the workers to "*procedan con prudencia por propia conveniencia abandonando pésimo sistema huelgas.*"

142. *El Imparcial,* August 2, 1912, 8.

143. *El Imparcial,* August 3, 1912, 1.

144. *El Imparcial,* August 2, 1912, 8.

145. *El Imparcial,* August 3, 1912, 1.

146. *El Imparcial,* August 5, 1912, 1 and August 6, 1912, 2. The newspaper reported (August 6) that weavers liked the new contract but that those in cloth preparation opposed it, suggesting the differences among mill hands.

147. *El Imparcial,* August 6, 1912, 1,2.

148. Ibid.

149. Ibid.

150. Ibid.

151. Ibid.

152. Ibid.

153. *El Imparcial,* August 8, 1912, 1 and 8, and August 9, 1912, 1 and 2.

154. Rivero Quijano, 1990, 359.

155. Rivero Quijano, 1990, 375.

156. Pánfilo Méndez to Félix Autran, July 31, 1911, AMRB, 1911.

157. El Alcalde Municipal to Presidente de la Sociedad Solidaridad Obrera, October 19, 1911, AMRB, 1911.

158. E.A.M.P.M.D.L.L. to Juez 1 de Paz, September 4, 1911, AMRB, 1911.

159. Sánches Vallejo to Presidente, February 10, 1912, AGN, DT, Caja 8, Exp. 4.

160. Ibid.

161. Ibid.

162. José González Soto y Hno. to Director del DT, August 10, 1912, AGN, DT, Caja 7, Exp. 28. "estos insensatos obreros . . . ."

163. Manuel Rivero Collado to Antonio Ramos Pedrueza, March 16, 1912, AGN, DT, Caja 7, Exp. 19.

164. JMC to Darío Mijares, Fábrica de San Bruno, March 15, 1912, AGEV, Sección de Fomento, Caja 27, Exp. 4s, 1912.

165. Darío Mijares to Secretario de Gobierno, Jalapa, March 15, 1912, AGEV, Sección de Fomento, Caja 27, Exp. 4s, 1912.

166. Miguel López Fuente to Director de la Oficina del Trabajo, March 9, 1912 AGN, DT, Caja 7, Exp. 11.

167. Quijano y Rivero to Director del Depto. del Trabajo, July 1, 1912, AGN, DT, Caja 7, Exp. 27.

168. Signoret y Reynaud to Señor Director del Departamento de Trabajo, May 10, 1912, AGN, DT, Caja 7, Exp. 13.

CHAPTER FIVE

1. Andrew Sayer, http://www.comp.lancs.ac.uk/sociology/papers/sayer-moral-economy.pdf., 2004.

2. J. Rivero Quijano, *La Industria Textil del Algodón y el Maquinismo en México* (Mexico, 1990), 375.

3. José González Soto y Hno. to director del Departamento del Trabajo, August 10, 1912, AGN, Caja 7, Exp. 28.

4. La Comición to Ramos Pedrueza, January 10, 1913, AGN, DT, Caja 49, Exp. 12.

5. For example, Julio Jiménez to Adalberto Esteva, April 8, 1913, AGN, DT., Caja 49, Exp. 30.

6. Leandro García to Antonio Ramos Pedrueza, December 24, 1912, AGN, DT, Caja 8, Exp. 26.

7. Leandro García to Antonio Ramos Pedrueza, December 24, 1912, AGN, DT, Caja 8, Exp. 26.

8. José Maria Alnaten to Antonio Ramos Pedrueza, January 9, 1913, AGN, DT, Caja 8, Exp. 26.

9. Manuel Ortega Elorza, June 14, 1913, AGN, DT, Caja 51, Exp. 15.

10. Departamento del Trabajo, Cuestionario para la estadística de diferencias y huelgas, December 1913, AGN, DT, Caja 1, Exp. 4.

11. Ibid.

12. Ibid.

13. Rámos Pedrueza to Zamacona, September 18, 1912, AGN, DT, Caja 7, Exp. 6.

14. Antonio de Zamacona to Señor Director, September 20, 1912, AGN, DT, Caja 7, Exp. 6.

15. Miguel Casas, Informe, February 10, 1913, AGN, DT, Caja 6, Exp. 27.

16. Admor. to Señor Jefe Político, January 7, 1913, AMA, Justicia y Seguridad Pública, Caja 337, 1913.

17. Antonio Ramos Pedrueza a los Obreros de la Fábrica, January 2, 1913, AGN, DT, Caja 47, Exp. 7.

18. Manuel R. Díaz to Román Rodríguez Peña, Subdirector del Departamento del Trabajo, March 24, 1913, AGN, DT, Caja 51, Exp. 1.

19. Manuel R. Díaz to Román Rodriguez Peña, Subdirector del Departamento del Trabajo, March 24, 1913, AGN, DT, Caja 51, Exp. 1.

20. Miguel Casas, "Informe que rinde . . . ," April 10, 1913, AGN, DT, Caja 51, Exp. 9.

21. Departamento de Trabajo, June 19, 1913, AGN, DT, Caja 37, Exp. 40.

22. Gabino Sánchez to Manuel Ortega Elorza, July 4, 1913, AGN, Caja 51, Exp. 145.

23. Isabel Hernández, et al., to Antonio Ramos Pedrueza, April 2, 1912, AGN, DT, Caja 7, Exp. 20.

24. Ma Trinidad Resendiz, Ma Rosario Yañez, Ma Feliza Lemus, and Ma Lina Marend to Sr. Antonio Ramos Pedrueza, August 3, 1912, AGN, DT, Caja 8, Exp. 16.

25. Miguel Casas to Adalberto Esteva, Director del Departamento del Trabajo, August 7, 1913, AGN, DT, Caja 51, Exp. 7.

26. Miguel Casas to Adalberto Esteva, August 7, 1913, AGN, DT, Caja 51, Exp. 7.

27. Manuel Ortega Elorza, September 23, 1913, AGN, DT, Caja 51, Exp. 14; see also Ortega Elorza to Adalberto Esteva, August 6, 1913, AGN, DT, Caja 51, Exp. 15.

28. Rivero, 1990, 317.

29. Internal memorandum, July 16, 1914, AGN, DT, Caja 86, Exp. 25.

30. Aguascalientes, Alberto Fuentes D., Decreto Num. 4, "La Limitación del Trabajo en bien de la clase obrera," AGN, DT, Caja 50, Exp. 29; Chiapas "Ley de Obreros," Periódico Oficial, October 30, 1914, AGN, DT, Caja 88, Exp. 1; Federal District, General Heriberto Jara, "a sus habitantes sabed . . . , September 28, 1914, AGN, DT, Caja 50, Exp. 29; Guanajuato, Abel Serrato, "Bases para . . . ," December 10, 1914, AGN, DT, Caja 50, Exp. 29; Jalisco, "Ley del Trabajo de Manuel Aguirre Berlanga," in R. Jorge Ortíz Escobar, *Legislación Laboral Veracruzana I* (Xalapa, 1999), 77–86; Michoacan, Gertrudis G. Sánchez, "Que deseando . . . ." October 28, 1914, AGN, DT, Caja 50, Exp. 29.

Puebla, Pablo González, "Hago Saber . . . ," September 2, 1914, AGN, DT, Caja 50, Exp. 29; Luis G. Cervantes, Considerando . . . , September 22, 1915, in Puebla, *Periódico Oficial,* vol. XCVI, no. 32, September 28, 1915; Luis G. Cervantes, Considerando . . . , December 21, 1915, *Periódico Oficial,* vol. XCVI, no. 44, December 21, 1915.

Tabasco, "Juntas Arbitrales del Trabajo Agricola," July 1913, AGEV, Fomento, Caja 40, Exp. 32; Luis F. Domínguez et al., "Decreto Relativo al Proletariado Rural," September 19, 1914, AGN, DT, Caja 88, Exp. 18; Tlaxcala, Máximo Rojas, "a sus habitantes sabed . . . ," September 7, 1914, AGN, DT, Caja 50, Exp. 29; Veracruz, Coronel Manuel Pérez Romero, "Considerando: que . . . , October 4, 1914, AGN, DT, Caja 50, Exp. 29; Cándido Aguilar, Núm. 8, October 13, 1914, AGN, DT, Caja 50, Exp. 29; Cándido Aguilar, Decreto Núm. 15, January 24, 1916, AGN, DT, Caja 88, Exp. 19; Yucatán, Eueuterio Avila, Decreto no. 4, September 11, 1914, AGN, DT, Caja, 50, Exp. 29.

31. Pablo González, "Hago Saber . . . ," September 2, 1914, AGN, DT, Caja 50, Exp. 29.

32. Secretario to Don Venustiano Carranza, September 29, 1914, AGN, DT, Caja 88, Exp. 16.

33. López Jiménez to Señor Director del Departamento del Trabajo, October 7, 1914, AGN, DT, Caja 88, Exp. 16.

34. The question of state versus federal labor law became a critical issue in the formation of the new labor regime, not to be settled until the 1931 Federal Labor Law, which repealed state law in this area. Clark, 1934, ch. 6.

35. Luis G. Cervantes, Considerando . . . , September 22, 1915, in Puebla, *Periódico Oficial,* vol. XCVI, no. 32, September 28, 1915.

36. Luis G. Cervantes, no. 44, December 21, 1915.

37. Marjorie Ruth Clark, *Organized Labor in Mexico* (Chapel Hill, 1934), 53.

38. Coronel Manuel Pérez Romero, "Considerando: que . . . ," October 4, 1914, AGN, DT, Caja 50, Exp. 29.

39. Cándido Aguilar, Decreto Número 11, October 19, 1914, AGN, DT, Caja 50, Exp. 29.

40. Cándido Aguilar, Decreto Número 15, January 24, 1916, AGN, DT, Caja 88, Exp. 19.

41. *El Dictamen,* January 27, 1916, 1.

42. To Oficial Mayor Encargado de la Secretaría de Fomento, October 2, 1914, AGN, DT, Caja 89, Exp. 10.

43. Venustiano Carranza, Decreto, December 1914, AGN, DT, Caja 89, Exp. 3.

44. Venustiano Carranza, "Decreto de Aumento de Jornales a los Obreros de la Industria Textil," March 22, 1915, in AGN, DT, Caja 50, Exp. 29.

45. Vicente Lombardo Toledano, *La Libertad Sindical en México* (Mexico, 1926), 53.

46. All of the contract articles are from Rafael Zubáran Capmany, *Proyecto de Ley sobre Contrato de Trabajo* (Veracruz, 1915).

47. Zubáran Capmany, 1915, 5.

48. Ibid., 6.

49. Ibid., 6.

50. A different view on the project can be found in José Villaseñor "Entre la Política y la Reivindicación," in Juan Felipe Leal and José Villaseñor, *La Clase Obrera en la Historia de México en la Revolución 1910–1917* Mexico, 1988), 346.

51. Martin Pérez, Julián Suárez, Luis Viveros to Antonio Valero, October 8, 1914, AGN, DT, Caja 87, Exp. 2.

52. Daniel Galindo to Director, Departamento del Trabajo, December 21, 1914, AGN, DT, Caja 91, Exp. 19.

53. José Natividad Díaz, et al., January 11, 1915, AGN, DT, Caja 104, Exp. 10.

54. José Natividad Díaz and Luis Viveros to Gerente de la Compañía Industrial de Orizaba S. A., February 8, 1915; Gerente to Sindicato, February 8, 1915; José Natividad Díaz and Luis Viveros to Marcos López Jiménez, February 12, 1915; AGN, DT, Caja 98, Exp. 2.

55. López Jiménez, Sub-director, to Señor Director del Departamento del Trabajo, October 7, 1914, AGN, DT, Caja 88, Exp. 16.

56. Secretario to Don Venustiano Carranza, September 29, 1914, AGN, DT, Caja 88, Exp. 16.

57. López Jiménez, Sub-director, to Señor Director del Departamento del Trabajo, October 7, 1914, AGN, DT, Caja 88, Exp. 16.

58. Antonio Valero, Director, to Oficial Mayor, October 9, 1914, AGN, DT, Caja 92, Exp. 6.

59. Acta del Sindicato, June 13, 1916, AGEV, DEPS, 1916, Exp. 703 (21-c).

60. Acta del Sindicato, June 14, 1916, AGEV, DEPS, 1916, Exp. 703 (21-c).

61. Acta, June 15, 1916, AGEV, DEPS, 1916, Exp. 703 (21-c).

62. Galindo and Heredia to Director, November 13, 1914, AGN, DT, Caja 91, Exp. 8.

63. Miguel Salazar et al. to Marcos López Jiménez, January 10, 1915, AGN, DT, Caja 97, Exp. 12.

64. AGN, DT, Caja 97, Exp. 12, Miguel Salazar et al. to Marcos López Jiménez, January 11, 1915.

65. Federico Hernández to José Taylor, April 22, 1915, AGN, DT, Caja 97, Exp. 14.

66. Ibid.

67. Joseph Taylor to Marcos López Jiménez, April 23, 1915, AGN, DT, Caja 97, Exp. 14.

68. Manuel Sánchez Martínez, et al. to Manuel R. Díaz, April 28, 1915, AGN, DT, Caja 97, Exp. 14.

69. Manuel Sánchez Martínez, et al. to Manuel R. Díaz, April 28, 1915, AGN, DT, Caja 97, Exp. 14.

70. Manuel R. Díaz to Marcos López Jiménez, May 8, 1915, AGN, DT, Caja 97, Exp. 14.

71. Manuel R. Díaz to Marcos López Jiménez, May 25, 1915, AGN, DT, Caja 97, Exp. 14.

72. Joseph Taylor to Director del Departamento del Trabajo, May 12, 1915, AGN, DT, Caja 97, Exp. 14.

73. Díaz to López Jiménez, May 25, 1915, AGN, DT, Caja 97, Exp. 14.

74. Macario Reyes to Manuel Díaz, March 14, 1915, AGN, DT, Caja 97, Exp. 30.

75. Macario Reyes to Manuel Díaz, March 15, 1915 AGN, DT, Caja 97, Exp. 30.

76. Enrique H. Hinojosa et al., March 15, 1915, AGN, DT, Caja 104, Exp. 11.

77. Daniel Galindo and Tomás Heredia to Director, November 13, 1914, AGN, DT, Caja 91, Exp. 8.

78. D. Thompson, February 13, 1915, AGN, DT, Caja 103, Exp. 8.

79. Daniel Galindo, June 24, 1915, AGN, DT, Caja 103, Exp. 8.

80. Daniel Galindo, Inspector, to Marcos López Jiménez, June 17, 1915, AGN, DT, Caja 107, Exp. 8.

81. Daniel Galindo to Marcos López Jiménez, June 17, 1915, AGN, DT, Caja 107, Exp. 8.

82. Memorandum, "En el orden en que . . . ," AGN, DT, Caja 104, Exp. 11.

83. Ibid.

84. John M. Hart, *Anarchism and the Mexican Working Class, 1860–1931* (Austin, 1987), chapter 10.

85. Gómez Alvarez, 1989, 49; *El Demócrata*, February 1, 1916, 5.

86. *El Demócrata*, November 3, 1916, 5.

87. *El Universal,* November 4, 1916, 1.
88. *El Universal,* November 6, 1916, p. 1; November 8, 1916, 1, 3.
89. *El Universal,* November 10, 1916, p. 1; November 11, 1916, 1.
90. *El Universal,* November 21, 1916, 1.
91. Asunto Toquero, July 11, 1916, ASRB, folder 1916.
92. El Dictamen, March 9, 1916, 4.

CHAPTER SIX

1. "Como se estableció la tarifa mínima de salarios para los obreros de hilados y tejidos de algodón," *Boletín del Departamento del Trabajo,* Año 1, no. 1, July 1913, 21.
2. Julio Gómez Abascal, Aviso a los Operarios, January 1, 1904, AMA, Presidencia, gobernación, 1904, Caja 91.
3. Helen L. Clagett and David M. Valderrama, *A Revised Guide to the Law and Legal Literature of Mexico,* Washington, 1973, 322.
4. "Proyecto de Ley," *Boletín del Departamento del Trabajo,* Año 1, no. 1, July 1913, 17.
5. "Iniciativa al Congreso de la Union para fundar el Departamento del Trabajo," *Boletín del Departamento del Trabajo,* Año 1, no. 1, July 1913, 17, 18.
6. StyPS, *Evolución Historica de la Secretaría del Trabajo y Previsión Social* (Mexico, 1957), 16.
7. In 1911 the Department began work with twelve employees and a budget of 46,317 pesos. By 1932, there were 276 employees and a budget of 900,664 pesos. "Iniciativa al Congreso . . . ," *Boletín del Departamento del Trabajo,* Año I, no. 1, July 1913, 18; StyPS, *Evolución Histórica de la Secretaría del Trabajo y Previsión Social* (Mexico, 1957), 62.
8. Antonio Zamacona, November 14, 1912, AGN, DT, Caja 3, Exp. 12.
9. Adalberto A. Esteva, "Labor del Departamento del Trabajo," *Boletín del Departamento del Trabajo,* Año I, no. 1, July 1913, 2.
10. March 29, 1913, AGN, DT, Caja 47, Exp. 7.
11. Jeffrey Bortz, "The Genesis of Mexico's Modern Labor Regime: The 1937–1939 Cotton Textile Convention," *The Americas* 52:1 (July 1995), 43–69.
12. Marjorie Ruth Clark, *Organized Labor in Mexico* (Chapel Hill, 1934), 21.
13. Tomás Reyes Retana, sesión del día July 8, 1912, AGN, DT, Caja 15, Exp. 18.
14. On the elections, Alan Knight, *The Mexican Revolution* (Lincoln, 1990), v. 2, 472.
15. There is a relatively large literature on the Congreso Constituyente, the Congress that drafted the constitution, and also a large legal literature on Mexican constitutional law. On the Constituyente, see E. V. Niemeyer Jr., *Revolution at Querétaro: The Mexican Constitutional Convention of 1916–1917* (Austin, 1974). Typical of the legal literature is J. Jésus Castorena, *Manual de Derecho Obrero* (Mexico, 1984). The best guide, to the time it was published, is Clagett and Valderrama, 1973. For Knight's judgment, Knight, 1990, V2, 469–478.

16. Clark, 1934, 45.

17. Lombardo declared that it was a "desarrollo fiel del principio filosófico de la Revolución francesa." Vicente Lombardo Toledano, *La Libertad Sindical en México* (Mexico, 1926), 13.

18. Alan Forrest, *The French Revolution* (Oxford, 1995), 86.

19. H. N. Branch, *The Mexican Constitution of 1917* compared with the Constitution of *1857* (Washington, 1926), 3.

20. Ernest Gruening, *Mexico and Its Heritage* (New York, 1928), 335.

21. Jesús Rivero Quijano, *La Revolución Industrial y La Industria Textil en México* (Mexico, 1990), 407.

22. *Boletín de la Secretaría de Gobernación,* "Constitución Política de los Estados Unidos Mexicanos," Mexico, October 1922, vol. 1, no. 5, p. 319.

23. Branch, 1926, 16.

24. I have used for this section *Constitución Política de los Estados Unidos Mexicanos, firmada el 31 de enero de 1917 y promulgada el 5 de febrero del mismo año* (Mexico, 1917), 38 pp.

25. Gruening, 1928, 336.

26. Yucatán, "Código del Trabajo del Estado de Yucatán, Decreto Núm. 722," Merida, July 28, 1917; "Ley del Trabajo expedida por el Gobierno Provisional de Tabasco el 14 de septiembre del presente Año", Villahermosa, 1917.

27. This is not a complete list.

28. Clark, 1934, 53; and Lombardo Toledano, 1926, 84, who commented that Veracruz's legislation "sirvió de modelo a otras." Veracruz had eighteen delegates to the Constituyente, and three of them played an important role in the development of labor affairs in Mexico, Adalaberto Tejeda, Heriberto Jara, and Aguilar.

29. *Ley del Trabajo del Estado Libre y Soberano de Veracruz-Llave* (Orizaba, 1918), 58.

30. Ibid., 59.

31. Ibid., 60.

32. Ibid., 62.

33. Author's underline.

34. Ibid., 78–79.

35. Ibid., 83.

36. Ibid., 83–84.

37. Ibid., 88–90.

38. Ibid., 93–95.

39. "La Junta Central, ejerciendo el arbitraje, es un tribunal de equidad y de justicia, no sujeto a los formulismos procesales ni a las tramitaciones largas y difíciles del derecho común." Ibid., 46.

40. Alfonso Cabrera, October 8, 1917, in *Periódico Oficial,* October 9, 1917, 233.

41. I have used the published and dedicated edition of December 1921, *Código de Trabajo* (Puebla, 1921). There is also the November 1921 version in the *Periódico Oficial.*

42. For example, the code notes that on profit sharing, it was necessary to "tomar experiencia de los acontecimientos surgidos en el Estado de Veracruz." *Código de Trabajo* (Puebla, 1921), x.

43. Ibid., 36.

44. Maurilio Luna et al., October 4, 1917, ASSR, 7, 1917, 714.1.

CHAPTER SEVEN

1. Jorge Basurto, *El Proletariado Industrial en México (1850–1930)* (Mexico, 1975).

2. Basurto, 1975, 62.

3. Juan Felipe Leal, *Del Mutualismo al Sindicalismo en México: 1843–1910* (Mexico, 1991), 60–61; Moisés González Navarro, *Las Huelgas Textiles en el Porfiriato* (Puebla, 1970), 13.

4. Felipe Leal, 1991, 96.

5. Marjorie Ruth Clark, *Organized Labor in Mexico* (Chapel Hill, 1934), 12.

6. Leticia Gamboa Ojeda, *Los Obreros Textiles de Puebla. El Caso de Atlixco, 1899–1924*, Tesis de Doctorado (Paris, 1993), 303.

7. David W. Walker, "Porfirian Labor Politics: Working Class Organization in Mexico City and Porfirio Díaz, 1876–1902," *The Americas*, vol. XXXVII, no. 3, January 1981, 282.

8. Gamboa, 1993, 323.

9. J. Rivero Quijano, *La Industria Textil del Algodón y el Maquinismo en México* (Mexico, 1990), 390.

10. Leandro Aguilar y Sánchez et al. to Ciudadano Jefe Político, November 15, 1907, AMA, Presidencia, Gobernación, 1907, Caja 164.

11. Pánfilo Méndez et al. to Jefe Político, December 26, 1908, ASRB, loose sheets, 1908–1920.

12. Ibid.

13. Manuel Toledano to Jefe Político, November 18, 1907, AMA, Presidencia, Gobernación, 1907, Caja 164.

14. Leandro Aguilar y Sánchez et al. to Ignacio Machorro, November 15, 1907, AMA, Presidencia, Gobernación, 1907, Caja 164.

15. Méndez continued in that position at least through 1909. Pánfilo Méndez to Jefe Politico, October 18, 1908, ASRB, folder 1912–1920; "En Río Blanco . . ." March 12, 1909, ASRB, loose sheets, 1908–1920.

16. *El Imparcial*, 1/14/12, p. 10.

17. Miguel Casas to Adalberto Esteva, March 1913, AGN, Caja 51, Exp. 14.

18. Clark, 1934, 17.

19. Ibid.

20. Ernest Gruening, *Mexico and Its Heritage* (New York, 1928), 336.

21. Obreros de la fábrica to Sr Director del departamento del trabajo, May 19, 1913, AGN, DT, Caja 48, Exp. 4.

22. Martín Pérez, Julian Suárez, Luis Viveros to Antonio Valero, October 8, 1914, AGN, DT, Caja 87, Exp. 2.

23. Victoriano García et al. to Gerente, January 23, 1915, AGN, DT, Caja 106, Exp. 26.

24. Alan Knight, *The Mexican Revolution* (Lincoln, 1990), v. 2, 424.

25. Antonio de Zamacona to Subdirector, May 27, 1912, AGN, DT, Caja 7, Exp. 13.

26. Ibid.

27. Elías Cano to Ramos Pedrueza, July 24, 1912, AGN, DT, Caja 7, Exp. 29.

28. Antonio de Zamacona, Estado en que se encuentra . . . , November 14, 1912, AGN, DT, Caja 3, Exp. 12.

29. Miguel Casas to Adalberto Esteva, June 1913, AGN, DT, Caja 51, Exp. 14.

30. Antonio de Zamacona to Director, September 20, 1912, AGN, DT, Caja 7, Exp. 6.

31. Ibid.

32. Quijano y Rivero to Rámos Pedrueza, June 28, 1912, AGN, DT, Caja 7, Exp. 25.

33. Director del Departamento to Nicolás Meléndez, March 15, 1912, AGN, DT, Caja 7, Exp. 19.

34. N. Melendez to A. Rámos Pedrueza, March 16, 1912, AGN, DT, Caja 7, Exp. 19.

35. Antonio Rámos Pedrueza to Manuel Rivero Collado, March 20, 1912, AGN, DT, Caja 7, Exp. 19.

36. Coralia Gutiérrez Álvarez, *Experiencias contrastadas. Industrialización y conflictos en los textiles del centro-oriente de México, 1884–1917* (Mexico, 2000), 271.

37. Ibid.

38. Antonio de Zamacona to Director, September 20, 1912, AGN, DT, Caja 7, Exp. 6.

39. A. Monroy, January 4, 1915, AGN, DT, Caja 105, Exp. 2.

40. Marcos López Jiménez, Exposición de motivos . . . , December 19, 1914, AGN, DT, Caja 89, Exp. 3.

41. Luis López and Tomás Norato to Director del Departamento de Legislacion y Trabajo, July 15, 1914, AGN, DT, Caja 86, Exp. 25.

42. [No author] July 16, 1914, AGN, DT, Cadge 86, Exp. 25.

43. Rosendo Salazar, *Las Pugnas de la Gleba* (Mexico, 1972), 37. See also John Mason Hart, *Revolutionary Mexico* (Berkeley, 1987), 114.

44. Hart, 1987, 107–11.

45. Barry Carr, *El Movimiento Obrero y la Política en México, 1910–1929* (México, 1991), 47.

46. El Sindicalista (Mexico City), Organo de los Sindicatos Constituidos en la "Casa del Obrero Mundial," Año 1, no. 10, Mexico, March 1, 1914, 1.

47. Ibid.

48. A. Monroy, January 4, 1915, AGN, DT, Caja 105, Exp. 9.

49. See, for example, José Natividad Díaz and Francisco de B. Salazar to Marcos López Jiménez, December 20, 1914, AGN, DT, Caja 91, Exp. 19.

50. Miguel Casas to Adalberto Esteva, n.d., AGN, Caja 51, Exp. 14.

51. Departamento del Trabajo, Indice de los Documentos que contiene el expediente núm. 47 formado con motivo de la huelga en la fábrica de hilados Río Grande Jalisco, AGN, DT, Caja 6, Exp. 27.

52. Antonio de Zamacona to Subdirector, May 27, 1912, AGN, DT, Caja 7, Exp. 13.

53. Rafael Silva et al. to Subdirector, May 6, 1912, AGN, DT, Caja7, Exp. 13.

54. Antonio de Zamacona to Director, September 20, 1912, AGN, DT, Caja 7, Exp. 6.

55. W. O. Jenkins to Antonio Ramos Pedrueza, March 26, 1912, AGN, DT, Caja 7, Exp. 20.

56. Miguel Casas to Adalberto Esteva, n.d., AGN, Caja 51, Exp. 14.

57. Ibid.

58. Director to Rafael Pérez Jiménez, Presidente del Comité Ejecutivo de Obreros, February 14, 1914, AGN, DT, Caja 91, Exp. 11.

59. The DT sent a notice to textile factories to forward lists of 5 percent of the workers in factories with less than 500 operarios, and 3 percent with more than 500. The DT would choose the factory delegates from among those on the lists, requiring that the worker currently serve in the factory, have been a textile worker for five years, and *"acreditar sus buenas constumbres con certificado de la autoridad local o del Jefe del a negociación en que trabaja,"* as well as *"saber leer, escribir y conocer las primeras reglas de aritimética."* Adalberto A. Esteva, "A los Obreros de las Fábricas de Hilados y Tejidos," April 30, 1913, AGN, DT, Caja, 47, Exp. 19.

60. Minuta de los Gobernadores, May 31, 1913, AGN, DT, Caja 50, Exp. 12. These labor boards would include *"10 obreros de distintos oficios,"* the state governor, the local Jefe Político, and *"1 profesor de Instrucción Primaria."* Personal de las Cámaras de Trabajo, May 30, 1913, AGN, DT, Caja 50, Exp. 12.

61. El Subsecretario to Secretario de Fomento, August 5, 1913, AGN, DT, Caja 50, Exp. 12.

62. Julio Poulat, "Proyecto de Exposición al Congreso para que la Secretaría se denomine de "Industria, Comercio y Trabajo." AGN, DT, Caja 50, Exp. 13.

63. Ibid.

64. Miguel G. Casas, Informe que rinde . . ., February 26, 1914; AGN, DT, Caja 91, Exp. 11.

65. Manuel Ortega Elorza to Adalberto A. Esteva, July 24, 1913, AGN, DT, Caja 51, Exp. 15.

66. *Corruption,* like many words, only has meaning within a social context. This is not the place for a long discussion of union corruption in a society like that of revolutionary Mexico, except to say that friends and enemies of unions often accused leaders of not following formal rules.

67. Andrés Cabrera to Adalberto A. Esteva, November 10, 1913, AGN, DT, Caja 37, Exp. 35.

68. Juan Cedillo to Director del Departamento de Trabajo, November 10, 1913, AGN, DT, Caja 37, Exp. 35.

69. Ibid.

70. Ibid.

71. El Director to Juan Cedillo, Juez Menor de Paz, Hacienda de Guadalupe, Atotonilco, Puebla, November 15, 1913, AGN, DT, Caja 37, Exp. 35.

72. El Subddirector Encargado del Despacho, August 4, 1914, AGN, DT, Caja 85, Exp. 8.

73. Marcos López Jiménez to Oficial Mayor, December 19, 1914, AGN, DT, Caja 89, Exp. 3.

74. Venustiano Carranza, Decreto, December 1914, AGN, DT, Caja 89, Exp. 3.

75. Enrique Hinojosa et al., January 13, 1915, AGN, DT, Caja 104, Exp. 6.

76. Pablo Gallardo, February 3, 1915, AMCM, legajo 7, 1914–1920, Exp. 7.

77. Porfiria Trujillo to Inspectores, July 1915, AGN, DT, Caja 104, Exp. 107.

78. Cristina Gómez Álvarez, *Puebla: los obreros textiles en la revolución 1911–1918*, (Puebla, 1989), 46; Entrevista celebrada, March 21, 1915, AGN, DT, Caja 106, Exp. 20.

79. Gómez Alvarez, 1989, 47.

80. Entrevista celebrada entre la Comisión del Centro Fabril de Orizaba y el C. Ministro de Gobernación, Lic. Rafael Zubarán Capmany, March 21, 1915, AGN, DT, Caja 106, Exp. 20.

81. Galindo and Heredia to Director, November 13, 1914, AGN, DT, Caja 91, Exp. 8.

82. Miguel G. Casas to Director, July 24, 1914, AGN, DT, Caja 91, Exp. 53.

83. Ibid.

84. El Sindicalista, Año 1, no. 1, Mexico, September 30, 1913.

85. Enrique H. Hinojosa et al., March 15, 1915, AGN, DT, Caja 104, Exp. 11.

86. Clark, 1934, 33.

87. Hart, 1987, 128.

88. Ibid., 183.

89. Ibid., 133.

90. Gruening, 1928, 338.

91. Article 123, Clause XVI, in Felipe Tena Ramírez, *Leyes Fundamentales de México* (1957), 870.

92. Some unions thus date their founding from the year of legal registration, though the union may have existed previously. For example, "En junio de 1917, los obreros de la fábrica Covadonga fundaron el primer sindicato en Puebla." Gómez Álvarez, 1989, 65.

93. Araiza, 1963, 161. There is some disagreement on this, Hart, 1987, 141; Clark, 1934, 35.

94. Luis Araiza, *Historia de la Casa del Obrero Mundial* (Mexico, 1963), 165.

95. Basurto, 1975, 185.

96. Clark, 1934, 57.

97. Basurto, 1975, 188.

98. Hart, 1987, 157.

99. Juan Lozano, Juan Anzúres, Andrés de León and Ricardo Treviño, a leader of the I.W.W. unions in the Tampico oil region.

100. Basurto, 1975, 190–95; Clark, 1934, 60–64; and Hart, 1987, 158.

101. Gregg Andrews, *Shoulder to Shoulder? The American Federation of Labor, the United States, and the Mexican Revolution 1910–1924*, Berkeley, 1991.

102. Ezequiel Salcedo, Celestino Gasca, Juan Rico, Ricardo Treviño, J. Marcos Tristán, Eduardo Moneda, Juan B. Fonseca, Fernando Rodarte, Juan Lozano, José López Cortés, Reynaldo Cervantes Torres, Adalberto Polo, Pedro Suárez, Pedro Rivera Flores, Salvador Alvarez, Samuel O. Yúdico, and José F. Gutiérrez.

103. Relative newcomers Cayetano Pérez Ruíz, Salustio Hernández, Carlos Gracidas and Robert Haberman later became important. The highly secretive Acción rarely counted more than 20 members, but they were the core of organized labor in Mexico. Clark, 1934, 63.

104. Samuel Malpica, *Atlixco: Historia de la Clase Obrera* (Puebla, 1989), 64.

105. Samuel Malpica, "Anarcosindicalismo o sindicalismo revolucionario en México (1906–1938)," in Jaime Tamayo and Patricia Valles, *anarquismo, socialismo y sindicalismo en las regiones* (Guadalajara, 1993), 95.

106. Samuel Malpica, *Atlixco: Historia de la Clase Obrera* (Puebla, 1989), 65; Leticia Gamboa Ojeda, "La CROM en Puebla y el Movimiento Obrero Textil en los Años 20," in *Memorias del Encuentro sobre Historia del Movimiento Obrero* (Puebla, 1984), 38; Malpica, 1993, 95.

107. Malpica, 1993, 95.

108. Gamboa Ojeda, 1984, 40.

109. Clark, 1934, 83.

110. Clark, 1934, 83.

111. Malpica, 1993, 96.

112. Gamboa, 1984, 49.

113. Leal, 1985, 97; Clark, 1934, 84.

CHAPTER EIGHT

1. Higinio Cosio and Ignacio Cardoso to Gobernador del Estado, May 28, 1918, ASSR, 6, 1918, Exp. 6250/21.

2. S. Gavito to Gobernador del Estado, December 22, 1921, AGN, DT, Caja 305, Exp. 8.

3. Maure to Presidente Municipal, September 12, 1918, AGEV, JCCA, 1918, Exp. 59.

4. Andrés Gines to Presidente Municipal, May 6, 1919, AMA, Presidencia, 1919.

5. *El Universal*, November 4, 1916, 1.

6. Gómez Haro and E. Artasanchy to Srio. de Industria, Comercio y Trabajo, February 27, 1920, AGN, DT, Caja 213, Exp. 33.

7. Ibid.

8. Ibid.

9. Ibid.

10. Memorandum. La Confederación Textil Presenta los Siguientes Asunto. n.d. AGN, DT, Caja 214, Exp. 6.

11. Presidente Muncipal to Secretario General de Gobierno, May 6, 1921, AMA, Presidencia, 1921, Caja 13.

12. Joaquín Salazar to Presidente Municipal, October 28, 1921, AMA, Presidencia, 1921, Caja 13.

13. Joaquín Salazar to Presidente Municipal, November 2, 1921, AMA, Presidencia, 1921, Caja 13.

14. Terrón, December 19, 1921, AMA, Presidencia, 1921, Caja H-12.

15. Pedro A. López et al. to Felipe Terrón, October 10, 1921, Caja 13, AMA, Presidencia, 1921.

16. Bernardo Barrolo to Presidente Municipal, April 21, 1922, AMA, Presidencia, 1922, Caja B-10, no. 1.

17. Bernardo Barrolo to Presidente Municipal, May 24, 1922, AMA, Presidencia, 1922, Caja B-10, no. 1.

18. Presidente Municipal Provisional to Secretario General de Gobierno, March 29, 1924, AMA, Presidencia, 1924, Caja H-21, no. 39.

19. "Estudio Preliminar . . . ," July 17, 1922, 3, AGN, DT, Caja 467, Exp. 2.

20. Presidente, Cámara de Industriales de Orizaba, to Juntas Central de Conciliación y Arbitraje del Estado, July 9, 1918, AGEV, JCCA, 1918, Exp. 37.

21. Avelino Gutiérrez to Presidente Municipal, June 26, 1920; Presidente Municipal to Secretario General del Gobierno, June 27, 1920, AMA, Presidencia, 1920, Caja H-3.

22. Leticia Gamboa Ojeda, "La CROM en Puebla y el Movimiento Obrero Textil en los Años 20," in *Memorias del Encuentro sobre Historia del Movimiento Obrero* (Puebla, 1984), 45.

23. Eduardo Vivanco, Presidente Municipal, to Administrador, June 1, 1919, AMA, Presidencia, 1919, AMA, Presidencia, 1919.

24. Presidente Municipal to Venustiano Carranza, July 12, 1919, AMA, Presidencia, 1919.

25. Administrador to Presidente del Ayuntamiento, January 5, 1922, AMA, Presidencia, 1922, Caja B-11, Exp. Huelgas.

26. Artasanchez and Gómez Haro to Ministro de Industria, Comercio y Trabajo, September 25, 1920, AGN, DT, Caja 305, Exp. 8.

27. Roberto Saviñón to Jefe del Departamento del Trabajo, January 2, 1922, AGN, DT, Caja 325, Exp. 12.

28. "Estudio Preliminar . . . ," July 17, 1922, 2, AGN, DT, Caja 467, Exp. 2.

29. Gómez Alvarez, 1989, 56–59.

30. Gómez Alvarez, 1989, 60–63.

31. González Casanova, 1980, 50.

32. Ibid., 52.

33. Jefe del Departamento to Alberto J. Pani, "El Conflicto Obrero de Puebla," May 10, 1918; AGN, DT, Caja 126, Exp. 5; El Conflicto Obrero de Puebla, AGN, DT, Caja 126, exp. 6.

34. Benigno Mata, August 12, 1918, AGEV, JCCA, 1918, legajo 3.

35. Edelmiro Maldonado, *Breve Historia del Movimiento Obrero* (Culiacán, 1981), 68.

36. Circular No. 10, Confederación Regional Obrera Mexicana, Luis N. Morones, ASSR 6, 1918, Exp. 603; various workers to Presidente del H. Ayuntamiento de Santa Rosa, September 17, 1918, AGEV, JCCA, 1918, Exp. 59.

37. Andrés Gines (Srio Gral sindicato) et al. to Presidente Municipal, April 5, 1919, AMA, Presidencia, 1919.

38. Eduardo Vivanco to Administrador, June 1, 1919, AMA, Presidencia, 1919.

39. Saúl Rodiles to Secretario de Estado y del Despacho, October 22, 1920, AGN, DT, Caja 213, Exp. 20.

40. Bernardo García Díaz, *Textiles del Valle de Orizaba (1880–1925)* (Jalapa,1990), 229.

41. Facundo Pérez Linares, "Los Años que Fueron Nuestros," *Los Días Eran Nuestros* (México, 1988), 170.

42. Pérez Linares, 1988, 171.

43. El León to Juez de Paz, August 14, 1920, AGN, DT, Caja 213, Exp. 18.

44. El León to Juez de Paz, August 14, DT, Caja 213, Exp. 16.

45. Ibid.

46. Bernardo Barrolo to Juez de Paz, August 14, 1920, AMA, Presidencia, 1920, Caja 413.

47. Miguel Flores, De cómo unos Hurtadores Hiceron Posible . . . ," *Los Dias Eran Nuestros* (Mexico, 1988), 241.

48. José de Jesús Salmerón to Presidente Municipal, June 21, 1920, AMA, Presidencia, 1920, Caja H-3.

49. Maure to Presidente Municipal, September 12, 1918, and the other letters in the expediente, AGEV, JCCA, 1918, Exp. 59.

50. Bernardo Barrolo to Presidente Municipal, n.d., AMA, Presidencia, 1921, Caja H-12.

51. Adalberto Monsalvo, Gonzalo García, and Agustín Arrazola, October 6, 1916, ASSR, 6, 1916, Exp. 508; Carta Circular a los Señores Empleados, May 16, 1917, ASSR, 6, 1917, Exp. 551.9.

52. Ernesto Cabrera Presidente Municipal, October 7, 1921, AMA, Presidencia, 1921, Caja 13.

53. Gustavo Hernández to Gobernador del Estado, October 12, 1920, AGN, DT, Caja 214, Exp. 6.

54. Catarino Gutiérrez et al., to Gobernador del Estado, October 20, 1920, AGN, DT, Caja 214, Exp. 6.

55. Ibid.

56. Bernardo Barrolo to Presidente Municipal, n.d., AMA, Presidencia, 1921, Caja H-12.

57. Juan Gutiérrez to Presidente Municipal, January 4, 1919, AMA, Presidencia, 1919.

58. El León to Juez de Paz, August 14, 1920, AGN, DT, Caja 213, Exp. 18.

59. García Díaz, 1990, 218.

60. Abraham Franco to Gobernador del Estado, March 19, 1918, AHEM, Trabajo e Industria, vol. 74, Exp. 51.

61. Vicente Díaz and Agustín Díaz to Gustavo Hernández, October 21, 1920; Eulalio Martínez et al. to Presidente de los Estados Unidos Mexicanos, December 17, 1920, AGN, DT, Caja 214, Exp. 6.

62. Manuel M. Conde Sucs and Ramón Golzarri to Centro Industrial Mexicano, October 22, 1920, AGN, DT, Caja 214, Exp. 6.

63. Gustavo Hernández and Vicente Díaz to Gobernador del Estado, October 23, 1920, AGN, DT, Caja 214, Exp. 6.

64. Acta, November 18, 1920, AGN, DT, Caja 214, Exp. 6.

65. Acta, January 8, 1921, AGN, DT, Caja 214, Exp. 6.

66. Epifanio Gómez to Esteban Flores, November 30, 1920, AGN, DT, Caja 214, Exp. 6.

67. Jesús Rivero Quijano to Srio. de Industria, Comercio, y Trabajo, December 13, 1920, AGN, DT, Caja 214, Exp. 6.

68. Memorandum, n.d., AGN, DT, Caja 214, Exp. 6.

69. "Huelgas," AMA, Presidencia, 1921, Caja H-12.

70. Bernardo Barrolo to Presidente Municipal, n.d., AMA, Presidencia, 1921, Caja H-12.

71. Ibid.

72. Bernardo Barrolo to Juez de Paz, August 14, 1920, AMA, Presidencia, 1920, Caja 413.

73. Francisco Sánches de Tagle to Jefe del Depto. del Trabajo, January 19, 1920, AGN, DT, Caja 222, Exp. 4.

74. Ibid.

75. Vda. de Francisco M. Conde to Presidente del Centro Industrial Mejicano, November 22, 1920, AGN, DT, Caja 213, Exp. 21.

76. Ibid.

77. Manuel M. Conde Sucs. to Dept. del Trabajo, March 27, 1920, AGN, DT, Caja 214, Exp. 5.

78. To Gobernador del Estado, March 31, 1920; Antonio Juncos to Director del Dept. del Trabajo, April 8, 1920, AGN, DT, Caja 214, Exp. 5.

79. Novoa to Jefe del Depto. Trabajo, July 31, 1920, AGN, DT, Caja 215, Exp. 12.

80. Roberto Saviñón to Jefe del Departamento del Trabajo, January 9, 1922, AGN, DT, Caja 325, Exp. 12.

81. Andrés Gines to Presidente Municipal, April 5, 1919, AMA, Presidencia, 1919.

CHAPTER NINE

1. F. Harvey Middleton, *Industrial Mexico 1919: Facts and Figures* (New York, 1919), vii.

2. Anita Brenner, *The Wind That Swept Mexico* (New York, 1943), 302 pp.

3. See, for example, Thomas Benjamin and Mark Wasserman, editors, *Provinces of the Revolution: Essays on Regional Mexican History, 1910–1929* (Albuquerque, 1990).

4. Ernest Gruening, *Mexico and Its Heritage* (New York, 1928), 335.

5. Similarly, it was the breakdown of tsarist authority during World War I that allowed the Bolshevik Revolution to succeed.

6. Unión de veteranos del trabajo, *Sucesos del Trabajo y sus Luchas de Antaño. Obra que narra los sufrimientos del trabajador desde 1800 a 1907* (Puebla, 1938), 111.

7. *El Imparcial*, December 28, 1911, 1.

8. Miguel López Fuente to Director de la Oficina del Trabajo, March 9, 1912, AGN, DT, Caja 7, Exp. 11.

9. "Durante muchos años . . . ," May 1914, AGN, DT, Caja 89, Exp. 5; Antonio de Zamacona to Director, September 20, 1912, AGN, DT, Caja 7, Exp. 6; and

José López Portillo y Rojas to Manuel Bonilla, December 8, 1912, AGN, DT, Caja 4, Exp. 11.

10. Barios Obreros to Director del Departamento de Trabajo, May 27, 1913, AGN, DT, Caja 52, Exp. 1.

11. Peyrot to Presidente de la Junta de Administración Civil, March 29, 1916, AMO, Caja 563.

12. Martín Pérez, Julian Suárez, Luis Viveros to Antonio Valero, October 8, 1914, AGN, DT, Caja 87, Exp. 2.

13. Comité Ejecutivo to Jefe del Departamento, October 2, 1916, AGEV, DEPS, 1916, Esp. 724, 92-s.

14. Gómez Haro and E. Artasanchy to Srio Industria, Comercio y Trabajo, February 27, 1920, AGN, DT, Caja 213, Exp. 33.

15. Cía. Industrial de S. Ant. Abad, S.A. to SICT, August 23, 1920, AGN, DT, Caja 207, Exp. 33.

16. Ibid.

17. Jesús Rivero Quijano, *La Revolución Industrial y La Industria Textil en México* (Mexico, 1990), 414.

18. Jeffrey Bortz and Marcos Aguila, "Earning a Living: A History of Real Wage Studies in 20th Century Mexico," *Latin American Research Review,* vol. 41, June 2006, 137.

19. Aurora Gómez-Galvarratio, "The Impact of Revolution: Business and Labor in the Mexican Textile Industry, Orizaba, Veracruz 1900–1930" (doctoral dissertation, Harvard University, 1999), 399.

20. Cited in Bortz and Aguila, 2006, 120.

21. Centro Industrial Mexicano, Reglamento Interior, December 3, 1906, in AGN, Gobernación, Legajos, Caja 817, Exp. 1.

22. Enrique Hinojosa, a leader of the Orizaba workers in 1915, remembered that "Anteriormente se acostumbraba trabajar por tiempo indefinido; se fijaba la hora de cinco a seis de la mañana, para salir a las nueve y media de la noche . . . "entrevista, March 21, 1915, AGN, DT, Caja 106, Exp. 20.

23. Antonio de Zamacona to Subdirector, May 31, 1912, AGN, DT, Caja 7, Exp. 14.

24. See, for example, Departamento de Trabajo, inspección de trabajo, Emilia G. Vda de Sta María and Maria H. Vda. de Goxxeza, AGN, DT, Caja 90, Exp. 6; Antonio de Zamacona to Subdirector, May 31, 1912, AGN, DT, Caja 7, Exp. 14.

25. To Antonio Ramos Pedrueza, January 8, 1913, AGN, DT, Caja 52, Exp. 1.

26. Pablo González, September 2, 1914, AGN, DT, Caja 50, Exp. 29.

27. Gertrudis G. Sánchez, "Que deseando . . . ," October 28, 1914, AGN, DT, Caja 50, Exp. 29.

28. Acta del Sindicato, June 14, 1916, AGEV, DEPS, Exp. 703 (21-c).

29. Memorandum. En el orden en que . . . , n.d., AGN, DT, Caja 104, Exp. 11.

30. Ibid.

31. SICT, cuestionario sobre trabajo, Río Blanco, June 6, 1921, AGN, DT, Caja 299, Exp. 1.

32. Srio de Gobierno to Presidente Municipal de Coatepec, March 20, 1923, AGEV, TPS, 1923 (3854), núm. 183.

33. Cándido Aguilar, Número 11, Artículo Séptimo, October 19, 1914, AGN, DT, Caja 50, Exp. 29.

34. *La Convención* (Mexico City), March 24, 1915, 2.

35. Puebla, *Código de Trabajo* (Puebla, 1921), 63–75.

36. "Como se estableció la tarifa mínima de salarios para los obreros de hilados y tejidos de algodón," *Boletín del Departamento del Trabajo,* Año 1, no. 1, July 1913, 22.

37. Memorandum. "En el orden en que . . . ," n.d., AGN, DT, Caja 104, Exp. 11.

38. *El Pueblo* (Mexico City), May 24, 1916, 1.

39. *Código de Trabajo* (Puebla, 1921), Articles 104, 110, 112.

40. "Convención Colectiva de Trabajo, Celebrada entre Industriales y Obreros del Industria Textil," México, Secretaría de Industria, Comercio y Trabajo, *La Industria, el Comercio y el Trabajo en México, Tomo III, del trabajo y la previsión social* (Mexico, 1928), 213.

41. Ibid., 223–25.

42. Ibid., 227–30.

43. Memorandum Núm. 1., n.d., AGN, DT, Caja 1159, Exp. 2.

44. Ibid.

45. Lagar to Srio. Industria, Comercio y Trabajo, August 27, 1927, AGN, DT, Caja 1159, Exp. 2.

46. Ibid.

47. Gregory S. Crider, "Reinventing and Institutionalizing Patronage after the Revolution: Workers and Sindicatos in the Textile Mills of Atlixco, Puebla, 1924–1940," paper presented to the Latin American Labor History Conference, April 1994.

48. Crider, 1994, 65.

49. Ibid., 66.

50. Gruening, 1928, 350.

51. A fine discussion is in Kevin J. Middlebrook, *The Paradox of Revolution Labor, the State, and Authoritarianism in Mexico* (Baltimore, 1995), 463 pp.

52. Los humildes servidores to Lic. Pedrueza, January 13, 1913, AGN, DT, Caja 52, Exp. 1.

53. José Natividad Díaz to Adalberto Esteva, April 25, 1913, AGN, DT, Caja 37, Exp. 40.

54. Ibid.

55. T. Jackson Lears, "The Concept of Cultural Hegemony," *American Historical Review, 90* (June 1985), 568.

56. There were guerrilla activities in Mexico but not on the scale of pre-1959 or Central America in the 1970s, and repression did not reach the level of the Southern Cone with their military governments.

Index